I0115083

TRANSACTIONS OF THE

AMERICAN PHILOSOPHICAL SOCIETY

HELD AT PHILADELPHIA

FOR PROMOTING USEFUL KNOWLEDGE

VOLUME 72, PART 2

A Review

of

Copepoda Associated with Sea Anemones and Anemone-like Forms (Cnidaria, Anthozoa)

ARTHUR G. HUMES

BOSTON UNIVERSITY MARINE PROGRAM
MARINE BIOLOGICAL LABORATORY
WOODS HOLE, MASSACHUSETTS

THE AMERICAN PHILOSOPHICAL SOCIETY

INDEPENDENCE SQUARE: PHILADELPHIA

1982

Copyright © 1982 by The American Philosophical Society

Library of Congress Catalog
Card Number 81-71035
International Standard Book Number 0-87169-722-X
US ISSN 0065-9746

INTRODUCTION

Associations of copepods with marine invertebrates are common, especially in tropical waters (Humes, 1970). The last twenty-five years have seen the discovery of a large number of new copepod associates. [The term associate is used here in the sense of Gooding (1957), referring to cases in which there is little definite evidence about the nature of the association.] At present the number of species of copepods known to be associated with marine invertebrates is estimated to be about 1,300.

This review encompasses all copepods, ranging from loosely associated to endoparasitic forms, which are associated with Actiniaria (sea anemones) and Corallimorpharia (anemone-like forms). For each of the 42 species of copepods information is given on the host, the site on the host when known, and the locality. New species are fully described. For known species brief notes are provided.

Included in this review are:

New taxa:
  Doridicola cylichnophorus sp. n.
    from *Heteractis crispa*
  Doridicola paterellis sp. n.
    from *Heteractis crispa*
  Doridicola scyphulanus sp. n.
    from *Heteractis crispa*
  Doridicola caelatus sp. n.
    from *Entacmaea quadricolor*
  Doridicola hispidulus sp. n.
    from *Entacmaea quadricolor*
  Doridicola penicillatus sp. n.
    from *Entacmaea quadricolor*
  Doridicola dunnae sp. n.
    from *Heteractis crispa*
  Doridicola titillans sp. n.
    from *Condylactis gigantea*
  Lambanetes stichodactylae gen. n., sp. n.
    from *Stichodactyla haddoni* and *Stichodactyla gigantea*
  Lambanetes gemmulatus gen. n., sp. n.
    from *Cryptodendrum adhaesivum*
  Metaxymolgus pertinax sp. n.
    from *Tealia coriacea* and *Tealia crassicornis*
  Metaxymolgus confinis sp. n.
    from *Anthopleura elegantissima*, *Anthopleura xanthogrammica*, and

1

Tealia piscivora
Metaxymolgus turmalis sp. n.
　from *Anthopleura artemisia*
Metaxymolgus sunnivae sp. n.
　from *Epiactis prolifera, Tealia crassicornis,* and *Tealia lofotensis*
Notoxynus crinitus sp. n.
　from *Heteractis crispa*
Verutipes laticeps gen. n., sp. n.
　from *Entacmaea quadricolor*
New host record:
　Doridicola magnificus (Humes, 1964)
　　from *Cryptodendrum adhaesivum*
New geographical records:
　Doridicola magnificus (Humes, 1964) and *Metaxymolgus cuspis* (Humes,
　　1964) from New Caledonia
　Asteropontius parvipalpus Stock, 1975a, from Jamaica, Bahamas, and
　　Puerto Rico
　Asteropontius longipalpus Stock, 1975a, from Jamaica. Male described for
　　first time.

METHODS OF COLLECTION AND STUDY

The actiniarians in New Caledonia were isolated in sea water in plastic bags immediately after collection in the field. The bags were emptied into pails containing sea water and a small amount of 95 per cent ethyl alcohol, sufficient to make approximately a 5 per cent solution, was added. After one or two hours the anemones were gently washed, rinsing the gastrovascular cavity as much as possible. The water was then poured through a fine net (120 holes per 2.5 cm, each opening about 100 μm) and the copepods recovered from the sediment retained. The copepods were then preserved in 70 per cent ethyl alcohol with two changes to avoid the precipitation of calcium sulphate.

The copepods sent to me from California were collected by Dr. Wim Vader, Dr. Sunniva Lönning and Mr. Gregory M. Ruiz.

The copepods were measured and dissected in lactic acid, using the wooden slide technique described by Humes and Gooding (1964). The setae on the caudal rami were not included in determining the body length. The lengths of the first antennal segments were measured along their posterior nonsetiferous margins.

The names of the host anemones belonging to the family Stichodactylidae are those used by Dunn (1981).

All figures were drawn with the aid of a camera lucida. The letter after the explanation of each figure refers to the scale at which it was drawn. The abbreviations used are: A_1 = first antenna, A_2 = second antenna, L = labrum, MX_2 = second maxilla, MXPD = maxilliped, and P_{1-4} = legs 1–4.

ACKNOWLEDGMENTS

The field work in New Caledonia during June–August 1971, was made possible by a grant (GB-8381X) from the National Science Foundation. Mr. Roger C. Halverson from the University of California at Santa Barbara aided in making the collections. The generous help provided by the staff of the Centre ORSTOM de Nouméa is acknowledged with special thanks. Such help included the loan of an automobile, the use of a gasoline-engined ship, the use of a skiff, and laboratory facilities.

The collection of the West Indian specimens in 1959 was made possible by a grant from the National Science Foundation. Dr. Richard U. Gooding provided valuable assistance during this field work.

The preparation of this review has been supported by a grant (DEB-77 11879) from the National Science Foundation.

I am especially indebted to Dr. Daphne F. Dunn, California Academy of Sciences, who identified the New Caledonian actiniarians, even though they were poorly preserved on account of difficulties in the field. This review would not have been possible in its present form without her help.

Dr. Charles E. Cutress, then at the Smithsonian Institution, identified the West Indian anemones collected by the author.

For the opportunity to study the Californian specimens I wish to thank Dr. Wim Vader, Tromsö Museum, Tromsö, Norway (then on sabbatical leave at the Bodega Marine Laboratory, Bodega Bay, California) and Mr. Gregory M. Ruiz, University of California at Santa Barbara. Through their generosity I have been able to include the Californian species in this review. I thank also Dr. Sunniva Lönning, University of Tromsö, who made collections of copepods from Californian sea anemones.

COPEPOD ASSOCIATES

Calanoida

Family Ridgewayiidae M. S. Wilson, 1958
Genus *Ridgewayia* Thompson and A. Scott, 1903

Ridgewayia fosshageni Humes and Smith, 1974

Host: *Bartholomea annulata* Lesueur.
Site: Forms aggregations in immediate vicinity of actiniarian.
Locality: Atlantic side of Panama (Humes and Smith, 1974).
Notes: In the laboratory the copepods form stable aggregations only near *Bartholomea annulata*, in contrast to unstable and dispersed aggregations around *Stichodactyla* (= *Stoichactis*) *helianthus* (Ellis) and rocks (Humes and Smith, 1974). Length of ♀ 0.75 mm, ♂ 0.68 mm.

Harpacticoida G. O. Sars, 1903

Family Laophontidae T. Scott, 1905
Genus *Laophonte* Philippi, 1840

Laophonte adamsiae Raibaut, 1966

Host: *Adamsia palliata* Bohadsch.
Site: On membranous folds (expansions of the pedal disk).
Locality: Sète, Mediterranean coast of France (Raibaut, 1966).
Notes: The copepods are confined to the membranous cuticula of the actiniarian, and are never found on the pagurid with which the actiniarian is associated (Raibaut, 1966). Length of ♀ 500 μm, ♂ 370 μm.

Poecilostomatoida Kabata, 1979

Family Lichomolgidae Kossmann, 1877
Genus *Aspidomolgus* Humes, 1969

Aspidomolgus stoichactinus Humes, 1969

Host: *Stichodactyla* (= *Stoichactis*) *helianthus* (Ellis) [= *S. anemone* (Ellis)].
Site: In washings; apparently living in gastrovascular cavity.
Localities: St. James, Barbados; South Bimini, Bahamas; near La Parguera, Puerto Rico; near Kingston, Jamaica (Humes, 1969). Curaçao; Bonaire; Barbados; near La Parguera, Puerto Rico; Marathon, Sombrero Key, Florida (Stock, 1975a).
Notes: Length of ♀ 1.84 mm, ♂ 1.60 mm.

Host: *Homosticanthus duerdeni* Carlgren (= *H. denticulosus* Lesueur).
Site: In washings.
Localities: Tobago Cays, Grenadines; St. Martin; near La Parguera, Puerto Rico (Stock, 1975a).

Genus *Doridicola* Leydig, 1853

Doridicola actiniae (Della Valle, 1880a)
= *Lichomolgus actiniae* Della Valle, 1880a
= *Lichomolgus anemoniae* Claus, 1889

Host: *Actinia cari* Delle Chiaje (= *Actinia concentrica* Risso, var. *viridis*). For synonymy see Andres (1883, p. 194), Carlgren (1949, p. 49), and Schmidt (1972, p. 67).
Site: Not given.
Locality: Naples, Italy (Della Valle, 1880a, 1880b).
Notes: Length of ♀ 1.9 mm, ♂ 1.5 mm.

Host: *Actinia equina* Linnaeus.
Site: Not given.
Locality: Banyuls, France (Stock, 1960; Carton, 1963; Bouligand, 1966).
Notes: Length of ♀ 1.5–2.3 mm, ♂ 1.5 mm (Stock, 1960).

Host: *Anemonia sulcata* (Pennant).
Site: On tentacles (Stock, 1966).
Localities: Banyuls, France (Stock, 1960; Bouligand, 1966; Carton, 1963); Roscoff, France (Carton, 1963); Plymouth, England (Briggs and Gotto, 1973); Strangford Lough, Northern Ireland (Briggs, 1973); Split, Jugoslavia (Stock, 1960); Trieste, Italy (Claus, 1889).
Citations: Zulueta (1911), Bouligand (1966).

Doridicola antheae Ridley, 1879
perhaps = *Paranthessius anemoniae* Claus, 1889
(see Humes and Stock, 1973, p. 81)

Host: *Anemonia sulcata* Pennant (= *Anthea cereus* Hertwig). For synonymy see Andres (1883, p. 198), Carlgren (1949, p. 50), and Schmidt (1972, p. 71).
Site: Tentacles.
Locality: Ilfracombe, North Devon, England (Ridley, 1879).
Notes: Length of ♀ 2 mm, ♂ unknown. The incomplete description of this species leaves its identity in doubt.

Doridicola cylichnophorus sp. n.
Figs. 1a–i, 2a–i, 3a–k

Type material.—18 ♀♀, 21 ♂♂ from one actiniarian, *Heteractis crispa* (Ehrenberg), in 3 m, north of Isle Maître, near Nouméa, New Caledonia, 22°19′30″S, 166°24′35″E, 13 July 1971. Holotype ♀, allotype, and 31 paratypes

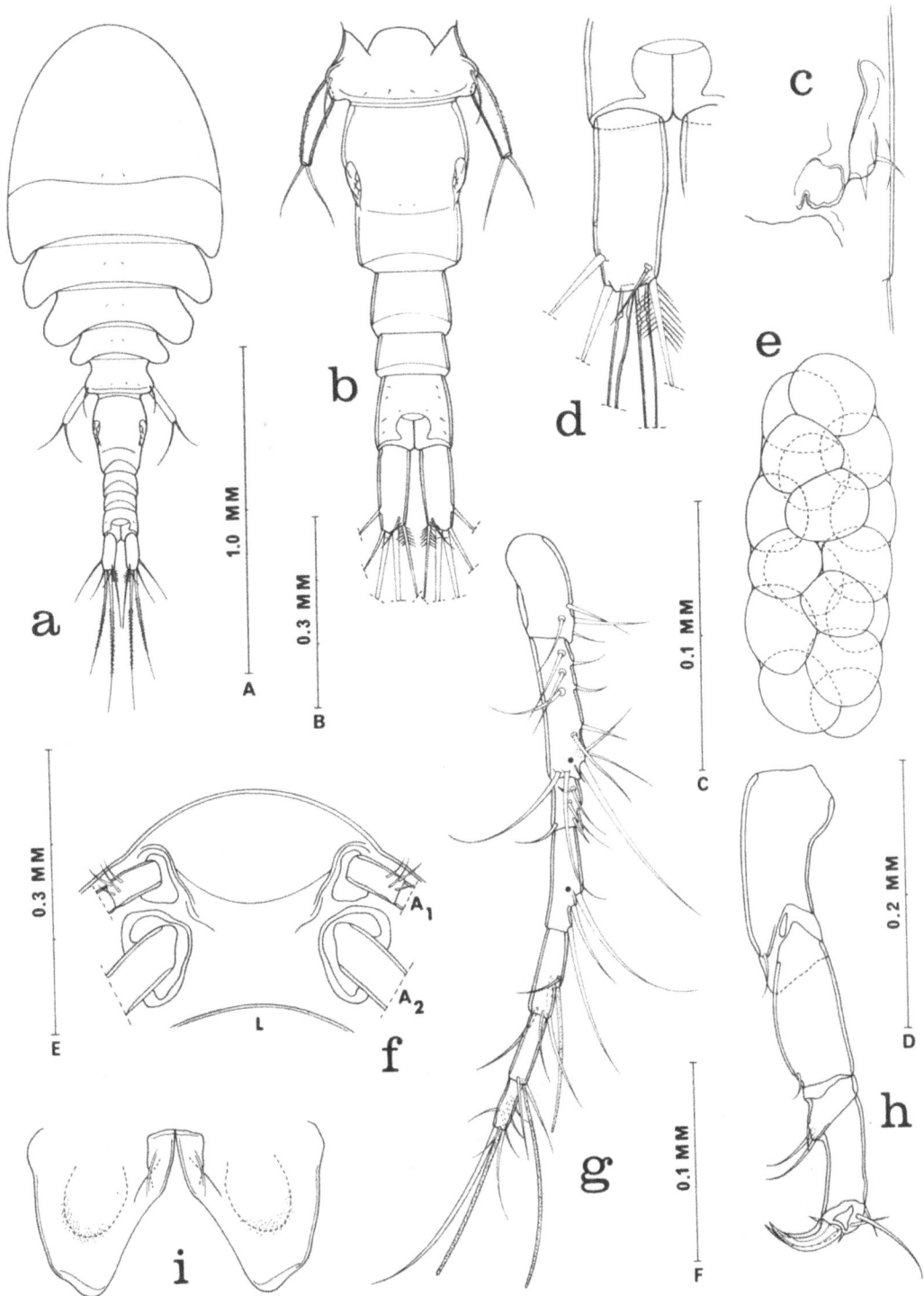

FIG. 1. *Doridicola cylichnophorus* sp. n., female: a, dorsal (A); b, urosome, dorsal (B); c, genital area, lateral (C); d, caudal ramus, dorsal (D); e, egg sac, dorsal (B); f, rostrum, ventral (E); g, first antenna, with dots indicating positions of aesthetes in male, dorsal (D); h, second antenna, anterior (D); i, labrum, with paragnaths indicated by broken lines, ventral (F).

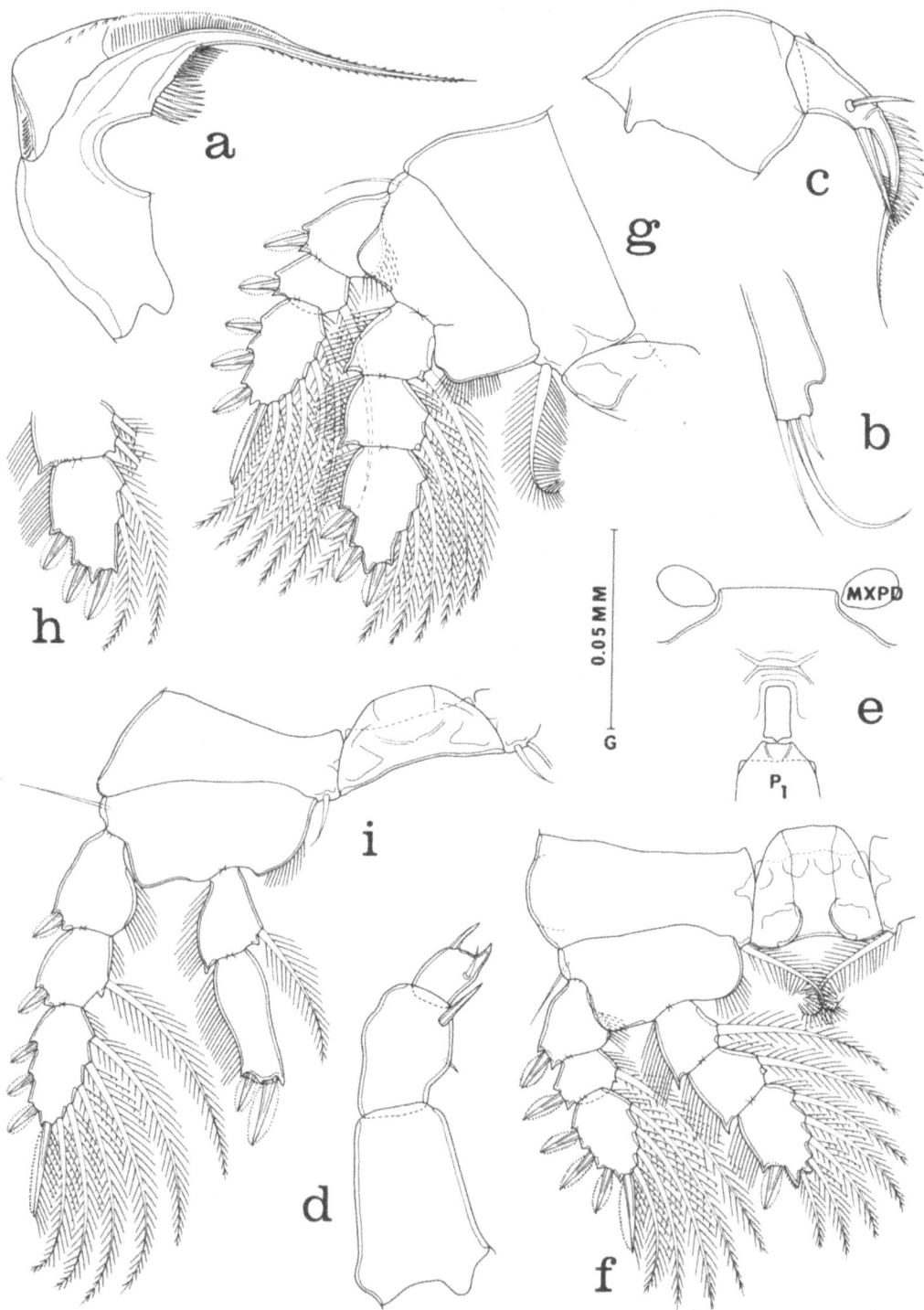

FIG. 2. *Doridicola cylichnophorus* sp. n., female: a, mandible, posterior (C); b, first maxilla, ventral (G); c, second maxilla, posterior (F); d, maxilliped, postero-inner (F); e, area between maxillipeds and first pair of legs, ventral (E); f, leg 1 and intercoxal plate, anterior (D); g, leg 2, anterior (G); h, third segment of endopod of leg 3, anterior (D); i, leg 4 and intercoxal plate, anterior (D).

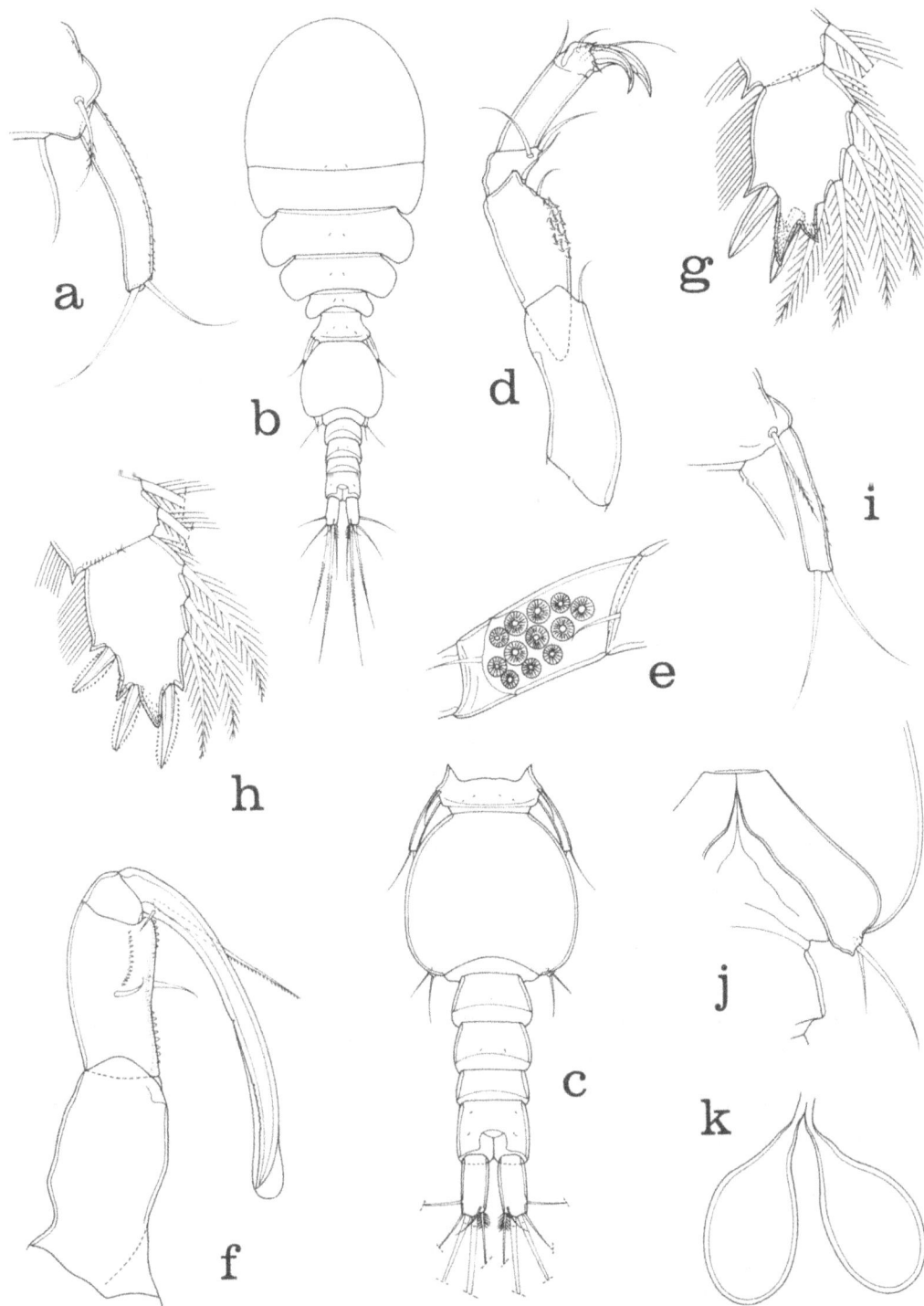

FIG. 3. *Doridicola cylichnophorus* sp. n. Female: a, leg 5, dorsal (D). Male: b, dorsal (A); c, urosome, dorsal (B); d, second antenna, posterior (D); e, second segment of second antenna, antero-inner surface (F); f, maxilliped, antero-outer (D): g, third segment of endopod of leg 1, anterior (F); h, third segment of endopod of leg 2, anterior (F); i, leg 5, dorsal (F); j, leg 6, ventral (D); k, spermatophores attached to female, ventral (E).

(14 ♀♀, 17♂♂) deposited in the National Museum of Natural History, Smithsonian Institution, Washington, D.C.; the remaining paratypes (dissected) in the collection of the author.

Other specimens (all from *Heteractis crispa*).—10 ♀♀, 8 ♂♂ from one host, in 0.5 m, western side of Isle To N'du, southwest of Pte. Laguerre, 10 km southwest of Paita, New Caledonia, 22°10′42″S, 166°16′30″E, 29 June 1971; 9 ♀♀, 7 ♂♂ from one host, in intertidal pool, southwestern corner of Port N'gea, 2 km north of Ricaudy Reef, near Nouméa, New Caledonia, 22°18′18″S, 166°26′47″E, 8 July 1971; 5 ♀♀, 12 ♂♂ from one host, in 15 cm, 5 km south of lagoon at Yaté, southeastern New Caledonia, 22°11′00″S, 166°59′00″E, 23 June 1971.

Female.—Body (Fig. 1a) with prosome not unusually broad. Length 1.68 mm (1.44–1.84 mm) and greatest width 0.67 (0.61–0.72 mm), based on 10 specimens. Segment of leg 1 separated from cephalosome by weak dorsal transverse furrow. Epimera of legs 1–4 rounded. Ratio of length to width of prosome 1.5:1. Ratio of length of prosome to that of urosome 1.6:1.

Segment of leg 5 (Fig. 1b) 117 × 218 μm. Genital segment longer than wide, in dorsal view 234 μm long, 174 μm wide in slightly broader anterior half, and 159 μm wide in somewhat narrower posterior half. Genital areas situated dorsolaterally near middle of segment. Each area (Fig. 1c) with two small naked setae 9 μm and 12 μm. Three postgenital segments from anterior to posterior 83 × 135, 75 × 127, 117 × 125 μm. Posteroventral border of anal segment with row of very small spinules on each side.

Caudal ramus (Fig. 1d) moderately elongate, 130 × 55 μm, ratio of length to width 2.36:1. Outer lateral seta 190 μm, dorsal seta 50 μm, and outermost terminal seta 151 μm, all three setae naked. Innermost terminal seta 224 μm with proximal hairs. Two long median terminal setae 385 μm (outer) and 473 μm (inner), both inserted between slight dorsal flange and small ventral flange with extremely minute marginal spinules. Both setae with very small lateral spinules along their midregions.

Body surface with few hairs (sensilla).

Egg sac (Fig. 1e) elongate, 594 × 231 μm, containing approximately 17 eggs with irregular shape, 104 × 135 μm in diameter.

Rostrum (Fig. 1f) broadly rounded.

First antenna (Fig. 1g) 428 μm long. Lengths of seven segments: 34 (78 μm along anterior margin), 104, 32, 70, 60, 48, and 36 μm respectively. Formula for armature: 4, 13, 6, 3, 4 + 1 aesthete, 2 + 1 aesthete, and 7 + 1 aesthete. All setae naked.

Second antenna (Fig. 1h) 4-segmented and 340 μm long including claws. Formula: 1, 1, 3, and 2 claws plus 5 setules. Fourth segment 99 μm along outer side, 59 μm along inner side, and 30 μm wide. Both claws about 57 μm. All setae smooth.

Labrum (Fig. 1i) with two rather slender posteroventral lobes. Mandible (Fig. 2a) with convex side of base having row of small spines on scalelike area followed by hyaline expansion and then serrated fringe preceded by few small spines. Concave side of base beyond identation bearing setules of two sizes. Lash moderately long and barbed. Paragnath (Fig. 1i) a small hairy lobe. First maxilla (Fig. 2b) with three smooth terminal setae and subterminal process. Second maxilla (Fig. 2c) 2-segmented, with unornamented first segment. Second segment with posterior surficial smooth seta, inner barbed spine, and outer proximal setule; lash long with graduated spines. Maxilliped (Fig. 2d) 3-segmented with unarmed first segment. Second segment with two unequal smooth setae. Third segment with two small naked setae and terminating in spiniform process.

Ventral area between maxillipeds and first pair of legs (Fig. 2e) only slightly protuberant.

Legs 1–4 (Fig. 2f–i) segmented and armed as follows (Roman numerals indicating spines, Arabic numerals representing setae):

P_1	coxa	0-1	basis	1-0	exp	I-0;	I-1;	III, I, 4
					enp	0-1;	0-1;	I, 5
P_2	coxa	0-1	basis	1-0	exp	I-0;	I-1;	III, I, 5
					enp	0-1;	0-2;	I, II, 3
P_3	coxa	0-1	basis	1-0	exp	I-0;	I-1;	III, I, 5
					enp	0-1;	0-2;	I, II, 2
P_4	coxa	0-1	basis	1-0	exp	I-0;	I-1;	III, I, 5
					enp	0-1;	II	

Coxa of leg 1 with slight outer posterior prominence. Leg 4 (Fig. 2i) with exopod 208 μm long. First endopod segment 55 × 49 μm, its inner plumose seta 96 μm. Second segment 86 μm long (91 μm including spiniform process), 39 μm in greatest width, and 31 μ in least width; two terminal fringed spines 26 μm (outer) and 47 μm (inner). Inner coxal seta of leg 4 32 μm and naked.

Leg 5 (Fig. 3a) with elongate free segment 131 × 26 μm, ratio 5:1. Two terminal setae smooth and 86 μm (inner) and 73 μ (outer). Segment ornamented with very small spinules along outer surface. Dorsal seta about 65 μm and delicately haired.

Leg 6 represented by two setae on genital area (Fig. 1c).

Color in living specimens in transmitted light pale brown, eye red, egg sacs gray.

Male.—Body (Fig. 3b) resembling in general form that of female. Length 1.46 mm (1.38–1.52 mm) and greatest width 0.53 mm (0.50–0.56 mm), based on 10 specimens. Ratio of length to width 1.58:1. Ratio of length of prosome to that of urosome 1.38:1.

Segment of leg 5 (Fig. 3c) 62 × 161 μm. Genital segment 247 × 260 μm, a little wider than long. Four postgenital segments from anterior to posterior 68 × 120, 65 × 117, 52 × 112, and 83 × 109 μm.

Caudal ramus (Fig. 3c) similar to that of female but shorter, 91 × 47 μm, ratio 1.94:1.

Rostrum as in female. First antenna like that of female but two aesthetes added (one on second segment and one on fourth segment, their positions indicated by dots in Figure 1g), so that formula is 4, 13 + 1 aesthete, 6, 3 + 1 aesthete, 4 + 1 aesthete, 2 + 1 aesthete, and 7 + 1 aesthete.

Second antenna (Fig. 3d) 300 μm long, resembling in general that of female but showing sexual dimorphism. Second segment with 12 saucer-shaped suckers on inner surface (Fig. 3e), each sucker 9–11 μm in diameter. Fourth segment 109 μm along outer side, 73 μm along inner side, and 26 μm wide. Two terminal claws 39 μm and 52 μm.

Labrum, mandible, paragnath, first maxilla, and second maxilla like those of female. Maxilliped (Fig. 3f) 4-segmented (assuming that proximal part of claw represents fourth segment). First segment unarmed. Second segment with two setae and two rows of spines. Small third segment unarmed. Claw 255 μm including terminal lamella, with slight indication of subdivision midway and bearing proximally two very unequal setae, the longer seta finely barbed distally along one edge.

Ventral area between maxillipeds and first pair of legs as in female.

Legs 1–4 segmented and armed as in female, except for third endopod segment of leg 1 where formula is I, I, 4 (Fig. 3g). Sexual dimorphism further evident in third endopod segment of leg 1 where two spines have smooth lamellae and where two terminal spiniform processes are large and minutely spinulose. Third endopod segment of leg 2 (Fig. 3h) with spines having serrate lamellae and two terminal processes also minutely spinulose. Legs 3 and 4 like those of female.

Leg 5 (Fig. 3i) with free segment 70 × 14 μm, ratio 5:1.

Leg 6 (Fig. 3j) a posteroventral flap on genital segment bearing two naked setae 42 μm and 81 μm.

Spermatophore (Fig. 3k) oval, 165 × 43 μm without neck.

Color in living specimens as in female.

Etymology.—The specific name *cylichnophorus*, from κυλιχνη = a small cup and φορεω = to bear, refers to the saucer-shaped suckers on the second antenna of the male.

Remarks.—Males of *Doridicola cylichnophorus* are notable in having suckers on the second segment of the second antenna. In *D. bulbipes* (Stock and Kleeton, 1963) and *D. magnificus* (Humes, 1964) this segment lacks fine ornamentation in the male. Although in several species of *Doridicola* this segment is ornamented [squamous in *Doridicola astrophyticus* (Humes and Stock, 1973), with spinules in *D. actiniae* (Della Valle, 1880a), *D. audens* (Humes, 1959), *D. foxi* (Gurney, 1927), *D. pteropadus* Humes, 1978, and *D. ptilosarci* Humes and Stock, 1973, with refractile knobs (*D. gemmatus* (Humes, 1964), with denticles (*D. isoawamochi* Ho, 1981), or with scalelike bosses (*D. virgulariae* Humes, 1978)], suckers on the second antenna have not previously been observed. It should be noted, however, that males are unknown in four species: *D. brevipes* (Shen and Lee, 1966), *D. buddhensis* (Thompson and A. Scott, 1903), *D. fishelsoni* (Stock, 1967), and *D. trispinosus* (Stock, 1959). In the remaining two species [*D. agilis* Leydig, 1853, and *D. rigidus* (Ummerkutty, 1962)] lack of information makes it impossible to determine whether or not sexual dimorphism in the fine ornamentation of the second antenna exists.

Females of the new species may be separated from most other species on the basis of the length of the caudal ramus. In 12 species the ratio is 1.5:1 or less (*D. agilis, D. audens, D. brevipes, D. buddhensis, D. bulbipes, D. foxi, D. gemmatus, D. pteropadus, D. ptilosarci, D. rigidus, D. trispinosus,* and *D. virgulariae*). In two species the ratio is 3.4:1 or more (*D. fishelsoni* and *D. magnificus*). In *D. astrophyticus* the length of the female is 1.30 mm, the free segment of leg 5 is ornamented with strong spines, and the inner seta on the first segment of the endopod of leg 4 is naked. The female of *Doridicola actiniae* (Della Valle, 1880a) as redescribed by Stock (1960), though similar in some respects to *Doridicola cylichnophorus,* differs from the New Caledonian species in several ways: the genital segment is distinctly broader in its anterior half, the fourth segment of the second antenna is relatively longer than in the new species, and the free segment of leg 5 is about 4:1 instead of 5:1 as in *D. cylichnophorus.*

Doridicola gemmatus (Humes, 1964)
= *Lichomolgus gemmatus* Humes, 1964

Host: *Stichodactyla gigantea* (Forskål) [= *Stoichactis giganteum* (Forskål)].
Site: In washings.
Locality: Region of Nosy Bé, northwestern Madagascar (Humes, 1964).
Notes: Length of ♀ 1.14 mm, ♂ 0.97 mm. Citation in Bouligand (1966).

Doridicola magnificus (Humes, 1964)
= *Lichomolgus magnificus* Humes, 1964

Host: *Stichodactyla gigantea* (Forskål) [= *Stoichactis giganteum* (Forskål)].
Site: In washings.
Localities: Region of Nosy Bé, northwestern Madagascar (Humes, 1964); near Nouméa, New Caledonia, new record, 1 ♂ from one host, intertidal, in sand, Ricaudy Reef, 22°19′00″S, 166°26′44″E, 21 July 1971.
Notes: Length of ♀ 3.06 mm, ♂ 2.74 mm. Citation in Bouligand (1966).

New host: *Cryptodendrum adhaesivum* (Klunzinger).
Site: In washings.
Localities: New Caledonia, new record, 1 ♀, 1 ♂ from one host, in 1 m, west of Isle Mando, near Nouméa, 22°18′59″S, 166°09′30″E, 3 July 1971. Citation in Bouligand (1966).
Remarks.—Additional information is given here to supplement the original description of the male. The rostrum (Fig. 4a) is rounded posteroventrally and bears many hairs. The lengths of the seven segments of the first antenna (Fig. 4b) are as follows: 99 (78 μm along anterior margin), 99, 50, 73, 55, 47, and 36 μm respectively. The total length is 459 μm. The third segment is apparently partly subdivided. All setae are smooth. The second antenna (Fig. 4c) is 365 μm long including the claws. The fourth segment is 108 μm along the outer side, 81 μm along the inner side, and 24 μm wide. The two terminal claws are 90 μm and 86 μm, both with a few minute teeth distally.

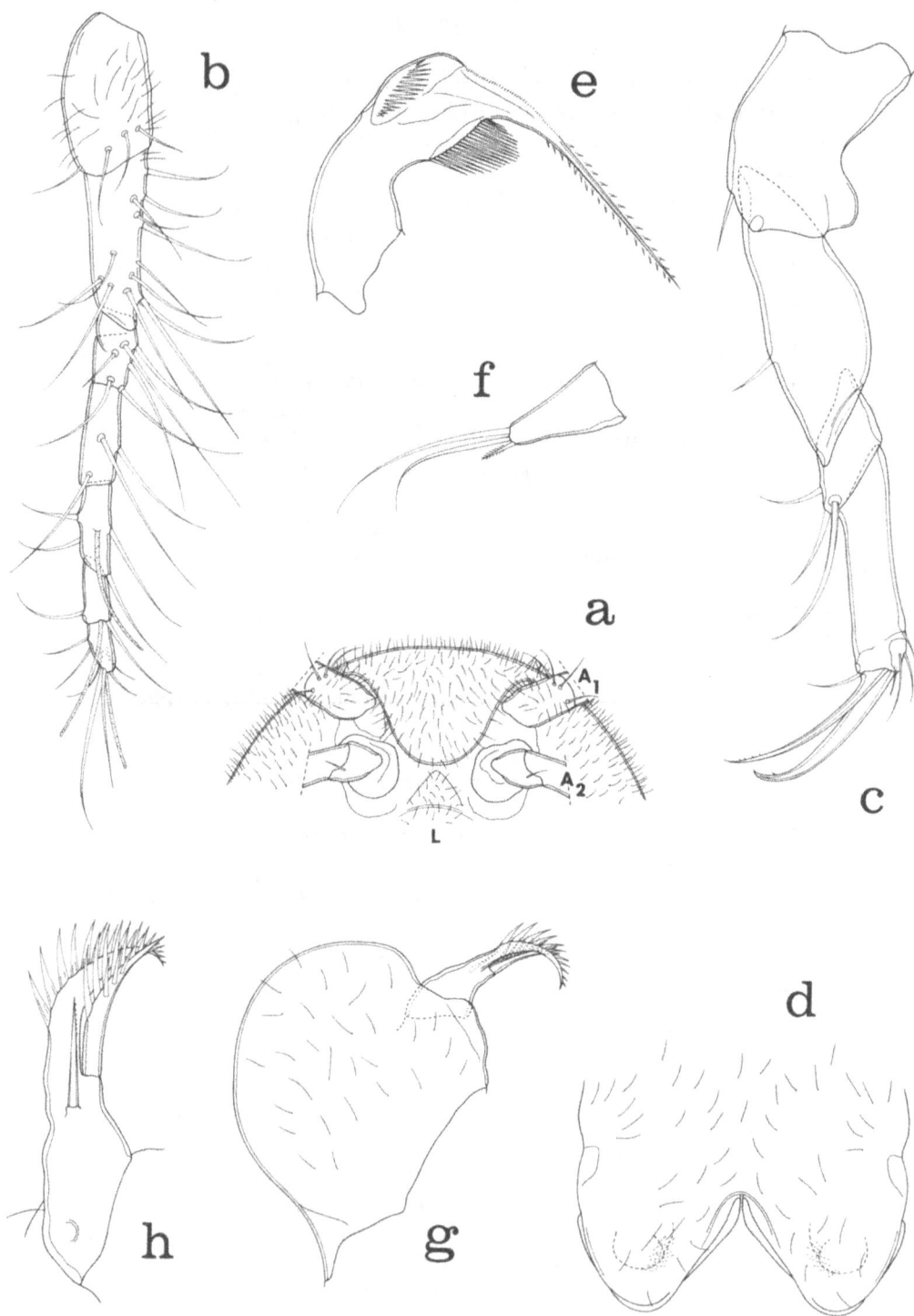

FIG. 4. *Doridicola magnificus* (Humes, 1964), male: a, rostrum, ventral (B); b, first antenna, anterior (F); c, second antenna, posterior (F); d, labrum, with paragnaths indicated by broken lines, ventral (B); e, mandible, posterior (C); f, first maxilla, anterior (C); g, second maxilla, anterior (F); h, second segment of second maxilla, posterior (G).

The labrum (Fig. 4d), mandible (Fig. 4e), paragnath (Fig. 4d), first maxilla (Fig. 4f), and second maxilla (Fig. 4g, h) do not differ greatly from those in the original description.

Doridicola paterellis sp. n.
Figs. 5a–k, 6a–j, 7a–h

Type material.—22 ♀♀, 40 ♂♂ from one sea anemone, *Heteractis crispa* (Ehrenberg), in 3 m, north of Isle Maître, near Nouméa, New Caledonia, 22°19′30″S, 166°24′35″E, 13 July 1971. (This is the same host individual from which *Doridicola cylichnophorus* was recovered.) Holotype ♀, allotype, and 54 paratypes (18 ♀♀, 36 ♂♂) deposited in the National Museum of Natural History, Smithsonian Institution, Washington, D.C.; the remaining paratypes (dissected) in the collection of the author.

Other specimens (all from *Heteractis crispa*): 33 ♀♀, 9 ♂♂ from one host, in 0.5 m, western side of Isle To N'du, southwest of Pte. Laguerre, 10 km southwest of Paita, New Caledonia, 22°10′42″S, 166°16′30″E, 29 June 1971; 8 ♀♀, 10 ♂♂ from one host, in intertidal pool, southwestern corner of Port N'gea, 2 km north of Ricaudy Reef, near Nouméa, New Caledonia, 22°18′18″S, 166°26′47″E, 8 July 1971; 4 ♂♂ from one host, in 15 cm, 5 km south of lagoon at Yaté, southeastern New Caledonia, 22°11′00″S, 166°59′00″E, 23 June 1971.

Female.—Body (Fig. 5a) moderately slender. Length 1.49 mm (1.36–1.62 mm) and greatest width 0.64 mm (0.60–0.69 mm), based on 10 specimens. Segment of leg 1 separated from cephalosome by indistinct dorsal furrow. Epimera of legs 1–4 rounded posteriorly. Ratio of length to width of prosome 1.77:1. Ratio of length of prosome to that of urosome 2.39:1.

Segment of leg 5 (Fig. 5b) 86 × 216 μm. Genital segment elongate, 200 × 159 μm, broadest in anterior third. Genital areas located dorsolaterally just anterior to middle of segment. Each area (Fig. 5c) with two small naked setae 10 μm and 16 μm. Three postgenital segments from anterior to posterior 73 × 99, 52 × 99, and 78 × 104 μm. Posteroventral border of anal segment with row of minute spinules on each side.

Caudal ramus (Fig. 5d) 88 μm long including ventral flange (83 μm with dorsal flange only) and 52 μm wide, ratio 1.69:1. Outer lateral seta 253 μm and naked. Dorsal seta 98 μm, outermost terminal seta 165 μm, innermost terminal seta 297 μm, and two long median terminal setae (inserted between smooth dorsal flange and ventral flange with row of minute marginal spinules) 418 μm(outer) and 462 μm(inner), all with lateral hairs or spinules.

Body surface with very few hairs (sensilla) (Fig. 5a).

Egg sac (Fig. 5e) elongate, 583 × 264 μm, containing approximately 22 eggs, 99–117 μm in diameter but of irregular form.

Rostrum (Fig. 5f) broad and shallowly rounded.

First antenna (Fig. 5g) 577 μm long. Lengths of seven segments: 52 μm (86 μm along anterior margin), 132, 36, 104, 88, 83, and 48 μm respectively. Formula for armature as in *Doridicola cylichnophorus.* All setae naked.

Second antenna (Fig. 5h) 4-segmented and 418 μm long including claws.

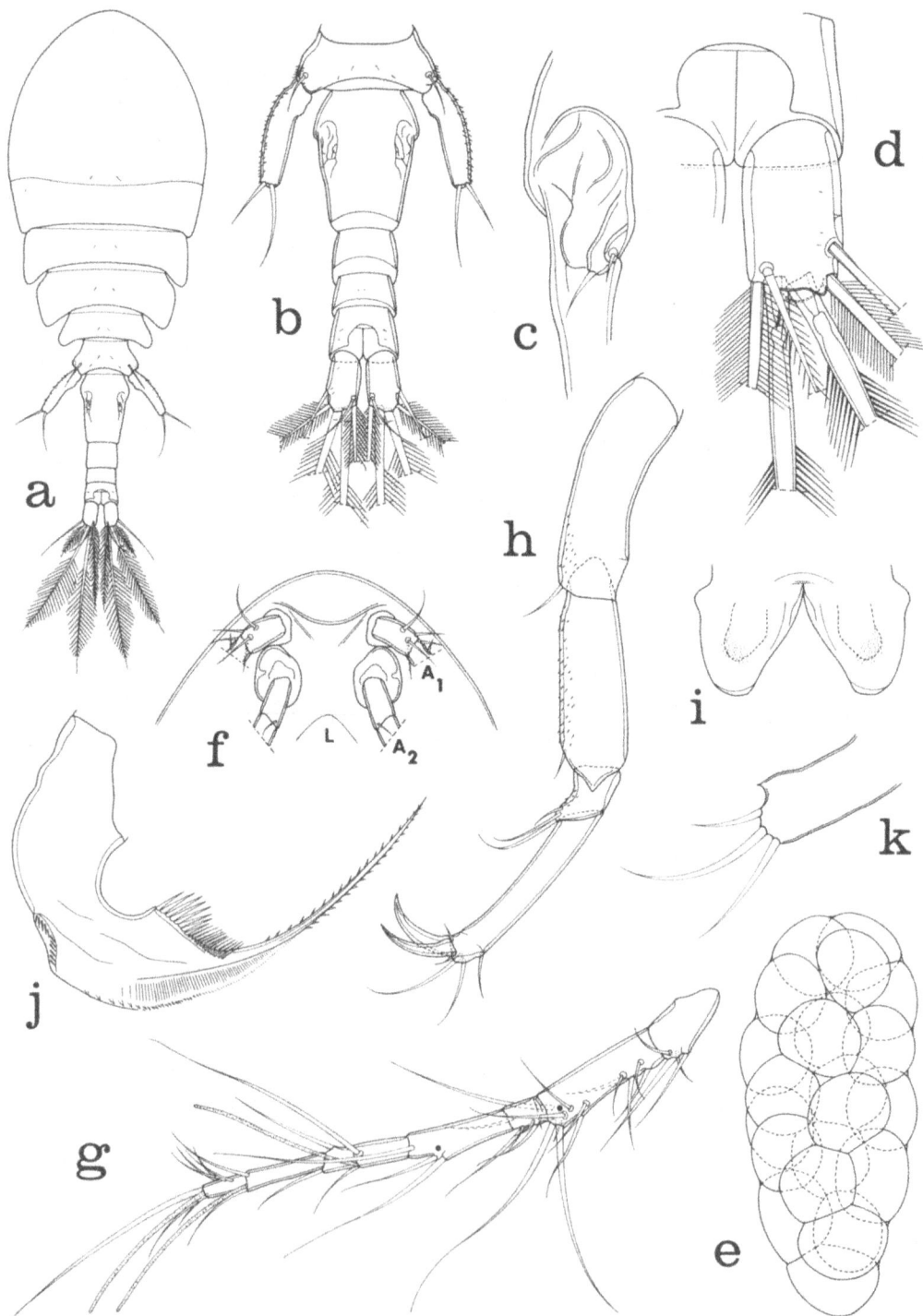

FIG. 5. *Doridicola paterellis* sp. n., female: a, dorsal (A); b, urosome, dorsal (B); c, genital area, dorsal (C); d, caudal ramus, dorsal (F); e, egg sac, ventral (B); f, rostrum, ventral (B); g, first antenna, with dots indicating positions of aesthetes in male, ventral (E); h, second antenna, posterior (D); i, labrum, with paragnaths indicated by broken lines, ventral (D); j, mandible, posterior (C); k, first maxilla, ventral (G).

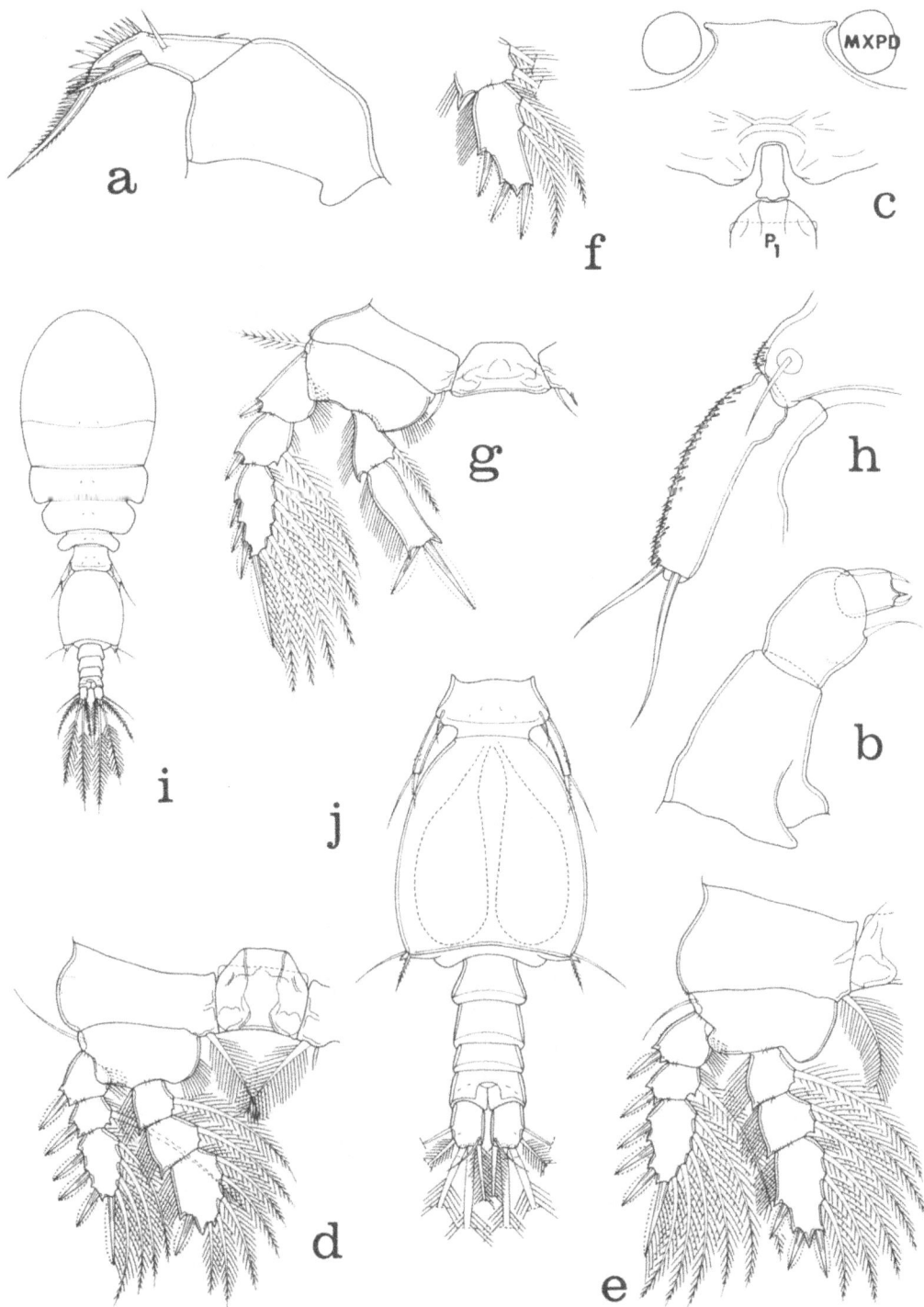

FIG. 6. *Doridicola paterellis* sp. n. Female: a, second maxilla, posterior (F); b, maxilliped, postero-inner (F); c, area between maxillipeds and first pair of legs, ventral (E); d, leg 1 and intercoxal plate, anterior (E); e, leg 2, anterior (E); f, third segment of endopod of leg 3, anterior (E); g, leg 4 and intercoxal plate, anterior (E); h, leg 5, dorsal (D). Male: i, dorsal (A); j, urosome, dorsal (E).

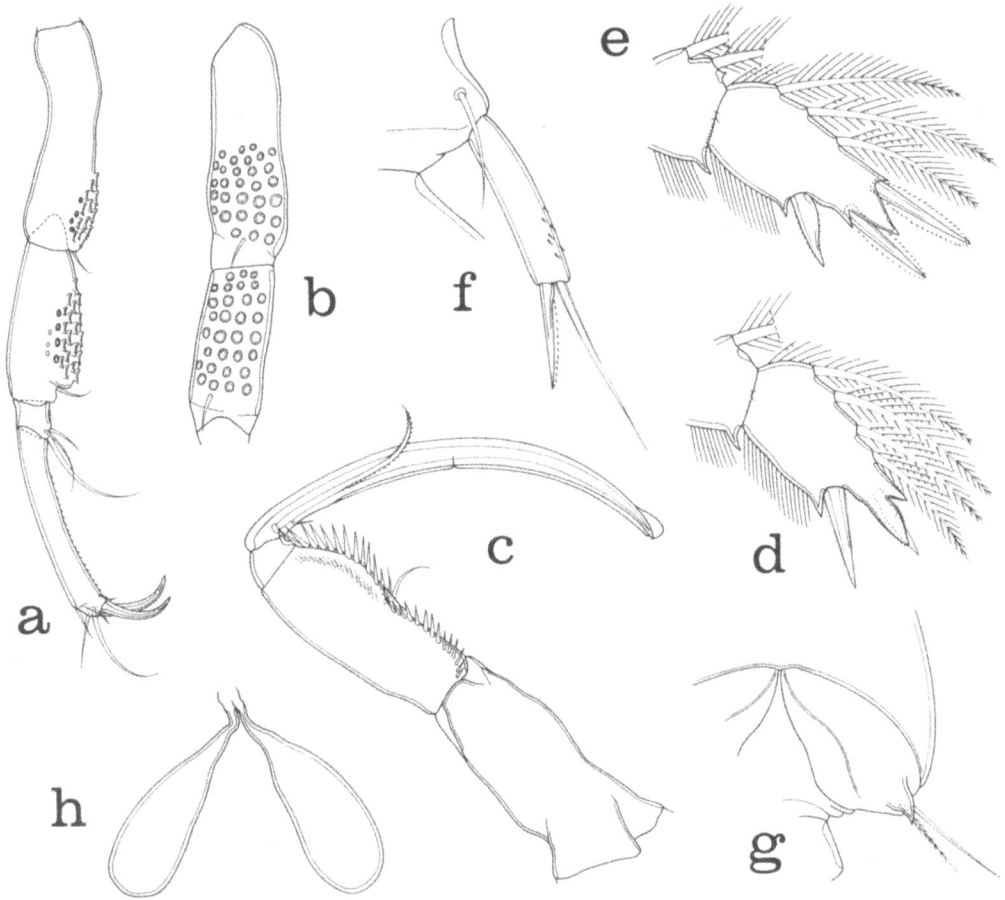

FIG. 7. *Doridicola paterellis* sp. n., male: a, second antenna, posterior (D); b, first and second segments of second antenna, inner (F); c, maxilliped, postero-inner (D); d, third segment of endopod of leg 1, anterior (F); e, third segment of endopod of leg 2, anterior (F); f, leg 5, dorsal (C); g, leg 6, ventral (D); h, spermatophores, attached to female, dorsal (E).

Formula as in *D. cylichnophorus*. All setae smooth. Two terminal claws slightly unequal, more slender claw 55 μm, stouter claw 60 μm. Fourth segment relatively long and slender, 156 μm along outer side, 117 μm along inner side, and 22 μm wide. Inner surface of first, second, and third segments with minute spinules.

Labrum (Fig. 5i) with two moderately slender posteroventral lobes. Mandible (Fig. 5j), paragnath (Fig. 5i), first maxilla (Fig. 5k), second maxilla (Fig. 6a), maxilliped (Fig. 6b), and ventral area between maxillipeds and first pair of legs (Fig. 6c) resembling those of *D. cylichnophorus*.

Legs 1–4 (Fig. 6d–g) segmented and armed as in *D. cylichnophorus*. Leg 4 with inner coxal seta 27 μm and naked. Exopod 208 μm. First segment of endopod 68 × 44 μm (without processes), its inner distal plumose seta

86 μm. Second segment of endopod 94 μm long without processes (104 μm with processes), 39 μm in greatest width, 29 μm in least width. Two terminal delicately fringed spines 52 μm (outer) and 87 μm (inner).

Leg 5 (Fig. 6h) with elongate free segment 161 × 42 μm (47 μm wide at slight inner proximal expansion), ratio 3.83:1, ornamented with small spines along outer side and armed with smooth terminal setae 81 μm and 112 μm. Adjacent dorsal seta approximately 65 μm and smooth. Outwardly near this seta a patch of small spines.

Leg 6 represented by two small setae on genital area (Fig. 5c).

Color in living specimens as in *D. cylichnophorus.*

Male.—Body (Fig. 6i) similar in general form to that of female. Length 1.13 mm (1.05–1.18 mm) and greatest width 0.42 mm (0.39–0.44 mm), based on 10 specimens. Ratio of length to width of prosome 1.70:1. Ratio of length of prosome to that of urosome 1.57:1.

Segment of leg 5 (Fig. 6j) 52 × 122 μm. Genital segment 234 × 199 μm, longer than wide and broadest in posterior half. Four postgenital segments from anterior to posterior 39 × 83, 42 × 79, 23 × 78, and 36 × 83 μm.

Caudal ramus resembling that of female but shorter 52 × 39 μm, ratio 1.33:1.

Rostrum as in female. First antenna similar to that of female but two aesthetes added (at positions indicated by dots in Figure 5g) as in *D. cylichnophorus.* Second antenna (Fig. 7a) 374 μm long, segmented and armed as in female. First two segments with inner suckers, approximately 17 on first segment and 26 on second segment (Fig. 7b). Third and fourth segments with small spinules along inner edges.

Labrum, mandible, paragnath, first maxilla, and second maxilla as in female. Maxilliped (Fig. 7c) resembling that of *D. cylichnophorus* but a little more slender than in that species. Claw 286 μm.

Ventral area between maxillipeds and first pair of legs as in female.

Legs 1–4 segmented and armed as in female except for third endopod segment of leg 1 (Fig. 7d) where formula is I, I, 4, with outer spine minutely fringed and inner spine smooth. Other sexual dimorphism seen in nature of terminal spiniform processes of third endopod segment of leg 1 (Fig. 7d) and leg 2 (Fig. 7e). Legs 3 and 4 like those of female.

Leg 5 (Fig. 7f) with free segment 61 × 13 μm, ratio 4.69:1, ornamented with only few small spines. Two terminal elements, inner a unilaterally fringed spine 36 μm and outer a smooth seta 65 μm. Smooth dorsal seta about 39 μm. Patch of small spines near dorsal seta absent.

Leg 6 (Fig. 7g) a posteroventral flap on genital segment bearing two setae, one 55 μm and feathered, other 88 μm and naked.

Spermatophore (Fig. 7h) elongate, approximately 185 × 68 μm without neck.

Color in living specimens as in female.

Etymology.—The specific name *paterellis* is formed from Latin *paterella* meaning a small cup or saucer and the suffix *-is* meaning provided with, referring to the suckers on the second antenna of the male.

Remarks.—As in *Doridicola cylichnophorus,* males of *Doridicola paterellis* may be distinguished from those of all other species in the genus by the presence of suckers on the second antenna. The male of *D. paterellis,* with suckers on both first and second segments of the second antenna, is readily separated from the male of *D. cylichnophorus,* with suckers only on the second segment of the second antenna.

Females of *D. paterellis* may be distinguished from those congeners having the ratio of length to width of the caudal ramus 1.5:1 or less, or 3.4:1 or more (see remarks on *D. cylichnophorus* above). The new species may be separated from females of the remaining species of *Doridicola* on other grounds. In *D. astrophyticus* the ratio of the caudal ramus is 2.13:1, and the second antenna lacks spinules on the first two segments and has a fourth segment shorter than in the new species. In *D. actiniae* the first maxilla has four setae, the ratio of the caudal ramus is 2.17:1, and the genital segment is more expanded laterally in the anterior half than in the new species.

Females of *Doridicola paterellis* may be distinguished from those of *D. cylichnophorus* by the relative length of the caudal ramus, the shape of the genital segment, the length of the fourth segment of the second antenna, and the nature of leg 5 (Table 1).

TABLE 1. Distinguishing features of *Doridicola paterellis* and *Doridicola cylichnophorus*

	D. paterellis	*D. cylichnophorus*
FEMALE		
Ratio of caudal ramus	1.69:1	2.36:1
Expanded part of genital segment	in anterior third	in anterior half
Ratio of greatest length to width of fourth segment of A_2	7:1	3.3:1
Leg 5	with slight inner proximal expansion	without inner proximal expansion
MALE		
Suckers on A_2	on both first and second segments	on second segment only

Doridicola scyphulanus sp. n.
Figs. 8a–m, 9a–o

Type material.—3 ♀♀, 14 ♂♂ from one sea anemone, *Heteractis crispa* (Ehrenberg), in 3 m, north of Isle Maître, near Nouméa, New Caledonia, 22°19′30″S,166°24′35″E, 13 July 1971. (This is the same host individual from which *Doridicola cylichnophorus* and *Doridicola paterellis* were recovered.)

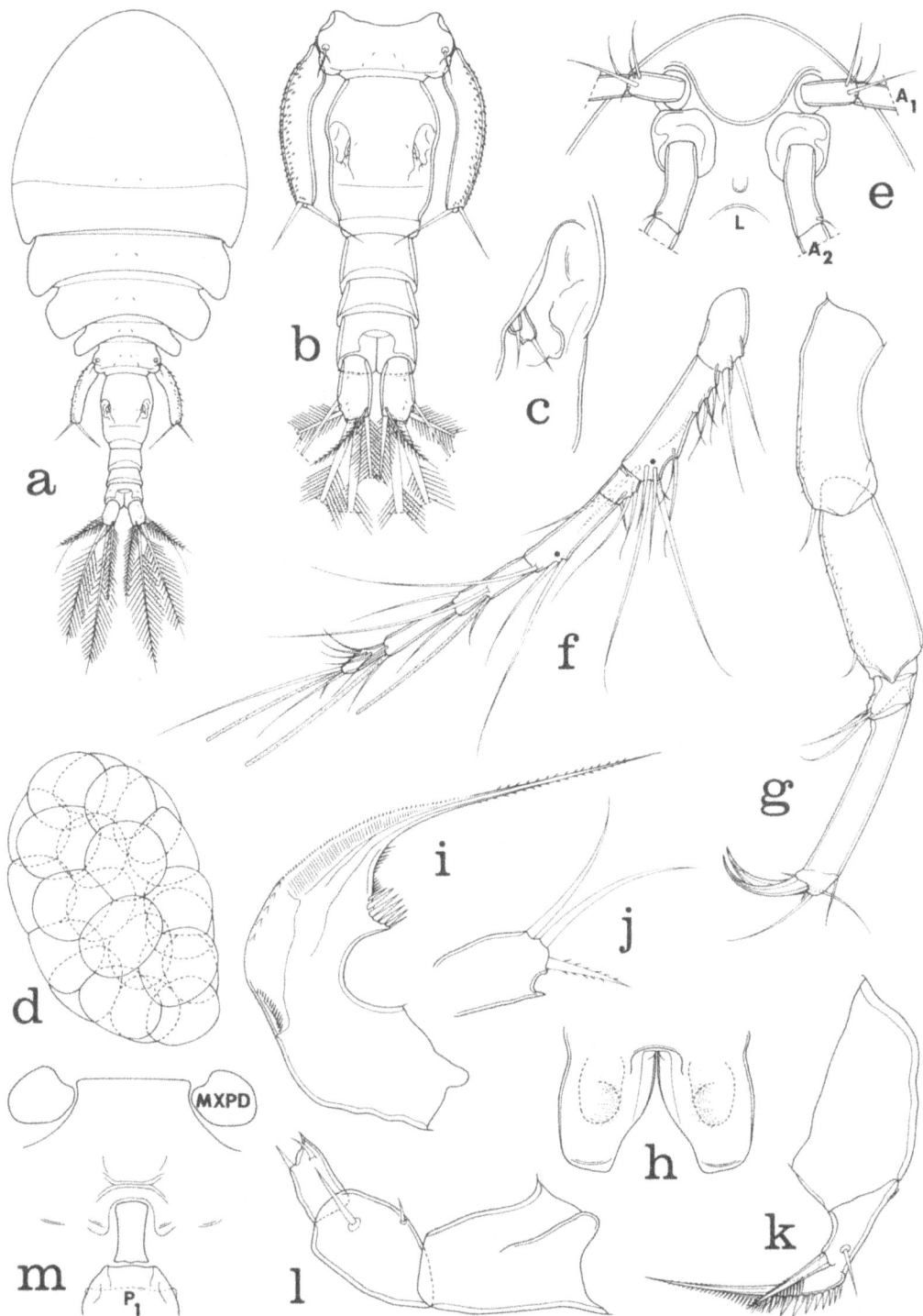

FIG. 8. *Doridicola scyphulanus* sp. n., female: a, dorsal (A); b, urosome, dorsal (B); c, genital area, dorsal (F); d, egg sac, ventral (B); e, rostrum, ventral (B); f, first antenna, with dots indicating positions of aesthetes in male, ventral (E): g, second antenna, posterior (D); h, labrum, with paragnaths indicated by broken lines, ventral (D); i, mandible, posterior (C); j, first maxilla, ventral (G); k, second maxilla, posterior (F); l, maxilliped, postero-inner (F); m, area between maxillipeds and first pair of legs, ventral (E).

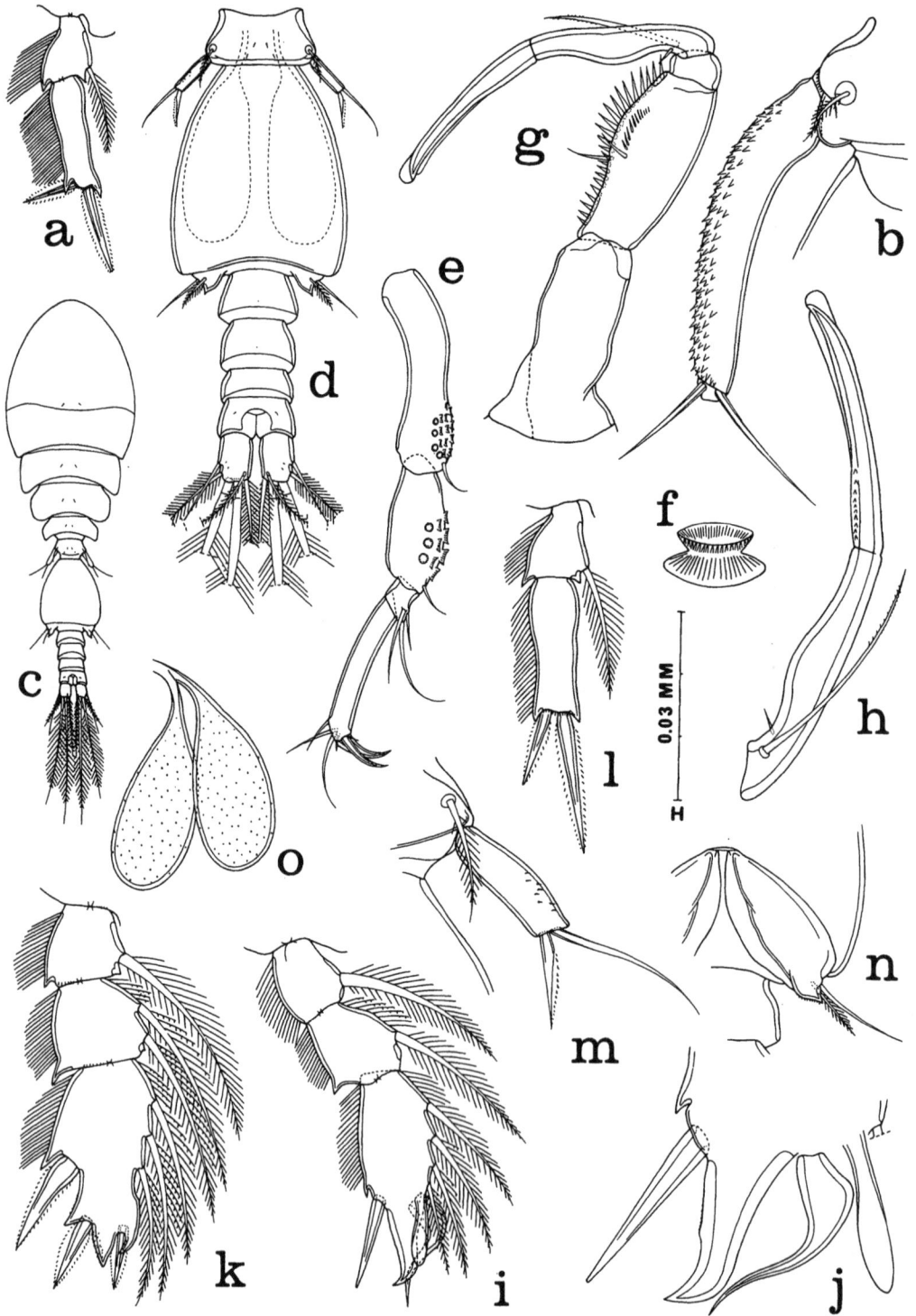

FIG. 9. *Doridicola scyphulanus* sp. n. Female: a, endopod of leg 4, anterior (E); b, leg 5, dorsal (D). Male: c, urosome, dorsal (A); d, urosome, dorsal (E); e, second antenna, posterior (D); f, sucker on second segment of second antenna, lateral (H); g, maxilliped, outer (D); h, claw of maxilliped, inner (F); i, endopod of leg 1, anterior (F); j, tip of endopod of leg 1, with elements separated, anterior (G); k, endopod of leg 2, anterior (F); l, endopod of leg 4, anterior (F); m, leg 5, dorsal (C); n, leg 6, ventral (D); o, spermatophores, attached to female, dorsal (E).

Holotype ♀, allotype, and 13 paratypes (2 ♀♀, 11 ♂♂) deposited in the National Museum of Natural History, Smithsonian Institution, Washington, D.C.; the remaining paratypes (dissected) in the collection of the author.

Other specimens (all from *Heteractis crispa*).—1 ♀ from one host, in 15 cm, in lagoon about 5 km south of Yaté, southeastern New Caledonia, 22°11′00″S,166°59′00″E, 23 June 1971; 2 ♀♀, 15 ♂♂ from one host, in 0.5 m, western side of Isle To N′du, southwest of Pte. Laguerre, 10 km southwest of Paita, New Caledonia, 22°10′42″S,166°16′30″E, 29 June 1971; 8 ♀♀ from one host, in tidal pool, southwestern corner of Port N′gea, 2 km north of Ricaudy Reef, near Nouméa, New Caledonia, 22°18′18″S, 166°26′47″E, 8 July 1971.

Female.—Body (Fig. 8a) with moderately broad prosome. Length 1.57 mm (1.46–1.66 mm) and greatest width 0.69 mm (0.65–0.74 mm), based on 10 specimens. Segment of leg 1 separated from cephalosome by weak dorsal furrow. Epimera of legs 1–4 rounded posteriorly. Ratio of length to width of prosome 1.34:1. Ratio of length of prosome to that of urosome 1.80:1.

Segment of leg 5 (Fig. 8b) 91 × 216 μm. Genital segment elongate, 220 × 169 μm in greatest dimensions, with sides nearly parallel. Genital areas located dorsolaterally near middle of segment. Each area (Fig. 8c) with two small naked setae about 13 μm long. Three postgenital segments from anterior to posterior 75 × 112, 44 × 112, and 78 × 108 μm. Posteroventral margin of anal segment with row of minute spinules on each side.

Caudal ramus (Fig. 8b) resembling that of *Doridicola paterellis*, 96 × 57 μm, ratio 1.68:1. Body surface (Fig. 8a) with very few hairs (sensilla). Egg sac (Fig. 8d) oval, 473 × 297 μm, containing 23–25 irregularly shaped eggs 100–138 μm in diameter.

Rostrum (Fig. 8e) broadly rounded.

First antenna (Fig. 8f) 587 μm long. Lengths of seven segments: 39 (91 μm along anterior margin), 133, 39, 101, 88, 86, and 49 μm respectively. Formula for armature as in *D. cylichnophorus* and *D. paterellis*. All setae naked.

Second antenna (Fig. 8g) 4-segmented and 423 μm long including claws. Formula as in two preceding species of *Doridicola*. All setae naked. Two terminal claws about 65 μm. Fourth segment long and slender, 156 μm along outer side, 120 μm along inner side, and 23 μm wide. Inner surface of first, second, and third segments with minute spinules.

Labrum (Fig. 8h), mandible (Fig. 8i), paragnath (Fig. 8h), first maxilla (Fig. 8j), second maxilla (Fig. 8k), maxilliped (Fig. 8l), and ventral area between maxillipeds and first pair of legs (Fig. 8m) resembling those of *D. paterellis*, differing only in small details.

Legs 1–3 as in *D. paterellis*. Leg 4 also resembling that of *D. paterellis* but endopod (Fig. 9a) slightly different. First segment of endopod 60 × 55 μm (length without spiniform processes), its inner plumose seta 82 μm (length of segment with spiniform processes 65 μm). Second segment of endopod 112 μm long without processes (127 μm with processes), 42 μm in greatest width, 31 μm in least width. Two terminal fringed spines 57 μm (outer) and 96 μm (inner), outer spine held at approximately a 90° angle to segment.

Leg 5 (Fig. 9b) with elongate free segment 234 × 39 μm (width taken at middle), ratio 6:1, carrying two smooth terminal setae 81 μm and 112 μm and ornamented with broad scalelike spines along outer side. Adjacent dorsal seta about 42 μm and finely plumose. Near this seta a row of very small spines.

Leg 6 represented by two small setae on genital area (Fig. 8c).

Color in living specimens as in *D. cylichnophorus* and *D. paterellis*.

Male.—Body (Fig. 9c) slender. Length 1.14 mm (0.95–1.12 mm) and greatest width 0.39 mm (0.33–0.43 mm), based on 10 specimens. Ratio of length to width of prosome 1.75:1. Ratio of length of prosome to that of urosome 1.49:1.

Segment of leg 5 (Fig. 9d) 55 × 122 μm. Genital segment 234 μm long including leg 6 (216 μm without leg 6) and 192 μm in greatest width, broadest in posterior third. Four postgenital segments from anterior to posterior 47 × 86, 47 × 83, 31 × 78, and 39 × 82 μm.

Caudal ramus resembling that of female but smaller, 52 × 39 μm, ratio 1.33:1.

Rostrum as in female. First antenna like that of female but two aesthetes added (at positions indicated by dots in Figure 8f) as in *D. cylichnophorus* and *D. paterellis*. Second antenna (Fig. 9e) 348 μm long, segmented and armed as in female. First two segments with inner suckers, approximately 16 on first segment with cup diameter about 5.5 μm and 10 on second segment with cup diameter 12 μm. Each sucker in lateral view (Fig. 9f) with broad hourglass shape.

Labrum, mandible, paragnath, first maxilla, and second maxilla like those of female. Maxilliped (Fig. 9g) resembling that in *D. paterellis* but claw 237 μm long including terminal lamella, distinctly swollen in proximal third and bearing row of small spines on inner surface just beyond subdivision (Fig. 9h).

Ventral area between maxillipeds and first pair of legs as in female.

Legs 1–4 segmented and armed as in female except for third endopod segment of leg 1 where formula as I, I, 4 (Fig. 9i), outer spine straight, inner spine swollen and angular proximally, both spines smooth. Between outer and inner spines a large hooklike spiniform process with few minute outer spinules. Medial to inner spine a spatulate smooth hyaline process. These two spines and two processes shown dissociated in Figure 9j. Other sexual dimorphism in terminal spiniform processes of third endopod segment of leg 2 (Fig. 9k). Leg 3 as in female. Leg 4 like that of female but outer spine on second segment of endopod (Fig. 9l) held nearly parallel to inner spine instead of at sharp angle as in female.

Leg 5 (Fig. 9m) with free segment 53 × 15 μm, ratio 3.53:1, its distal dorsal surface ornamented with few small spines. Two terminal elements, inner unilaterally fringed spine 35 μm and outer smooth seta 60 μm.

Leg 6 (Fig. 9n) a posteroventral flap on genital segment bearing two setae, one 55 μm and feathered, other 75 μm and smooth.

Spermatophore (Fig. 9o) elongate, about 195 × 86 μm not including long neck. Surface of spermatophore with many small refractile points.

Color in living specimens as in female.

Etymology.—The specific name *scyphulanus*, from Latin *scyphulus* meaning a little cup and the suffix *-anus* signifying belonging to or pertaining to, alludes to the suckers on the second antenna of the male.

Remarks.—*Doridicola scyphulanus* has many features in common with *Doridicola paterellis* and like that species *D. scyphulanus* may be distinguished from its congeners as described above.

The chief points of distinction between the new species and *D. paterellis* are summarized in Table 2.

TABLE 2. Distinguishing features of *Doridicola scyphulanus* and *Doridicola paterellis*

	D. scyphulanus	D. paterellis
FEMALE		
Shape of genital segment	sides nearly parallel; very slight hourglass form	broadest in anterior third tapering posteriorly
Egg sac	oval	elongate
Leg 5	about as long as genital segment, ratio 6:1	shorter than genital segment, ratio 3.83:1
MALE		
A$_2$	without spinules on inner margin of fourth segment	with spinules along inner margin of fourth segment
Maxilliped	claw distinctly swollen proximally, with row of small spines near middle	claw not swollen proximally, without row of small spines
P$_1$, third segment of endopod	inner spine swollen and angular proximally; outer terminal process stout and hooked with very few minute outer spinules	inner spine straight; outer terminal process straight and covered with minute spinules

Doridicola caelatus sp. n.
Figs. 10a–l, 11a–l, 12a–g

Type material.—24 ♀♀, 14 ♂♂ from five individuals of the actiniarian *Entacmaea quadricolor* (Rueppell and Leuckart), intertidal, eastern end of Isle Maître, near Nouméa, New Caledonia, 22°20′35″S, 166°25′10″E, 16 July 1971. Holotype ♀, allotype, and 31 paratypes (20 ♀♀, 11 ♂♂) deposited in the National Museum of Natural History, Smithsonian Institution, Washington, D.C.; the remaining paratypes (dissected) in the collection of the author.

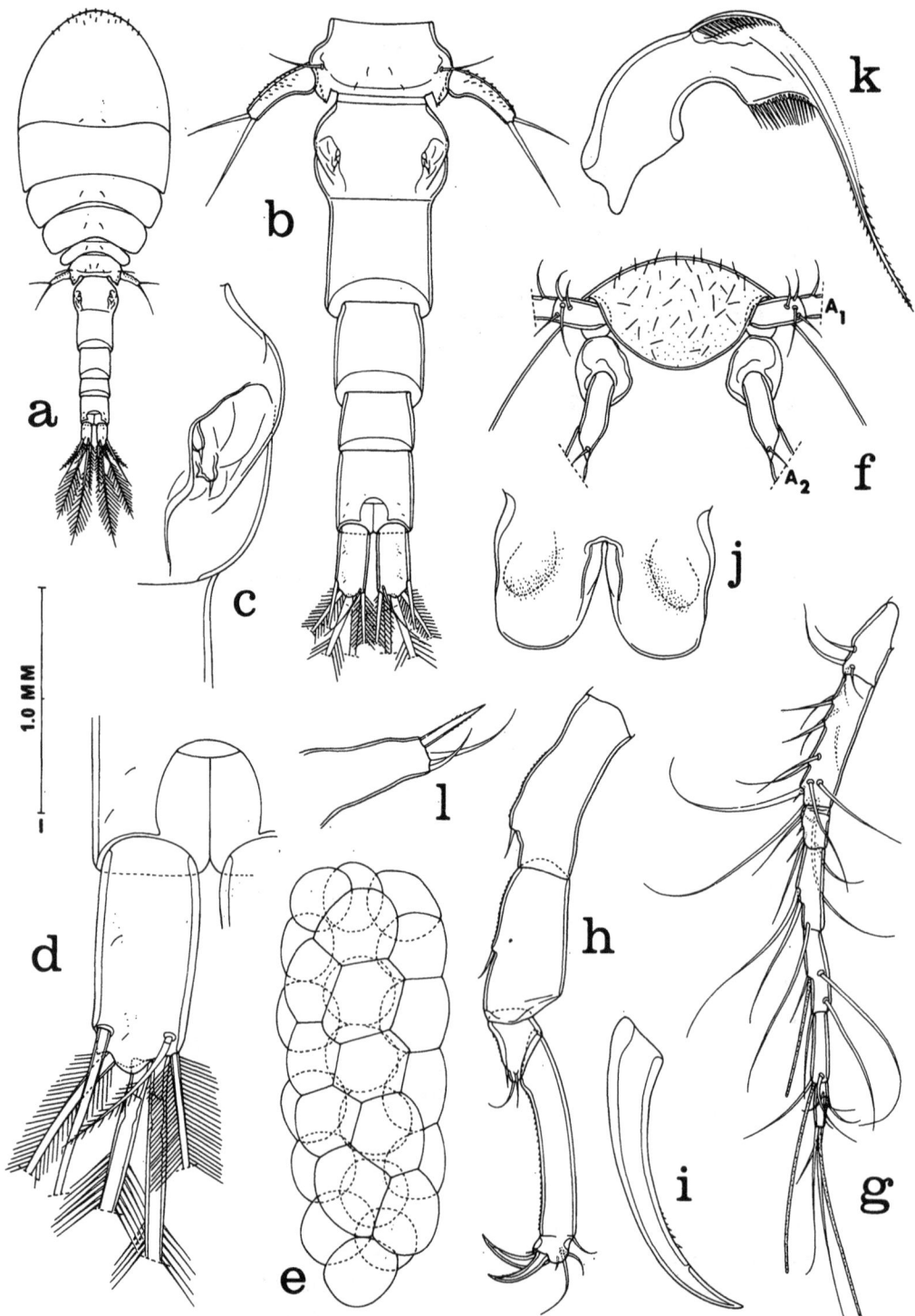

FIG. 10. *Doridicola caelatus* sp. n., female. a, dorsal (I); b, urosome, dorsal (B); c, genital area, dorsal (F); d, caudal ramus, dorsal (F); e, egg sac, ventral (B); f, rostrum, ventral (B); g, first antenna, ventral (E); h, second antenna, anterior (D); i, slender claw on last segment of second antenna, anterior (H); j, labrum, with paragnaths indicated by broken lines, ventral (F); k, mandible, posterior (C); l, first maxilla, ventral (G).

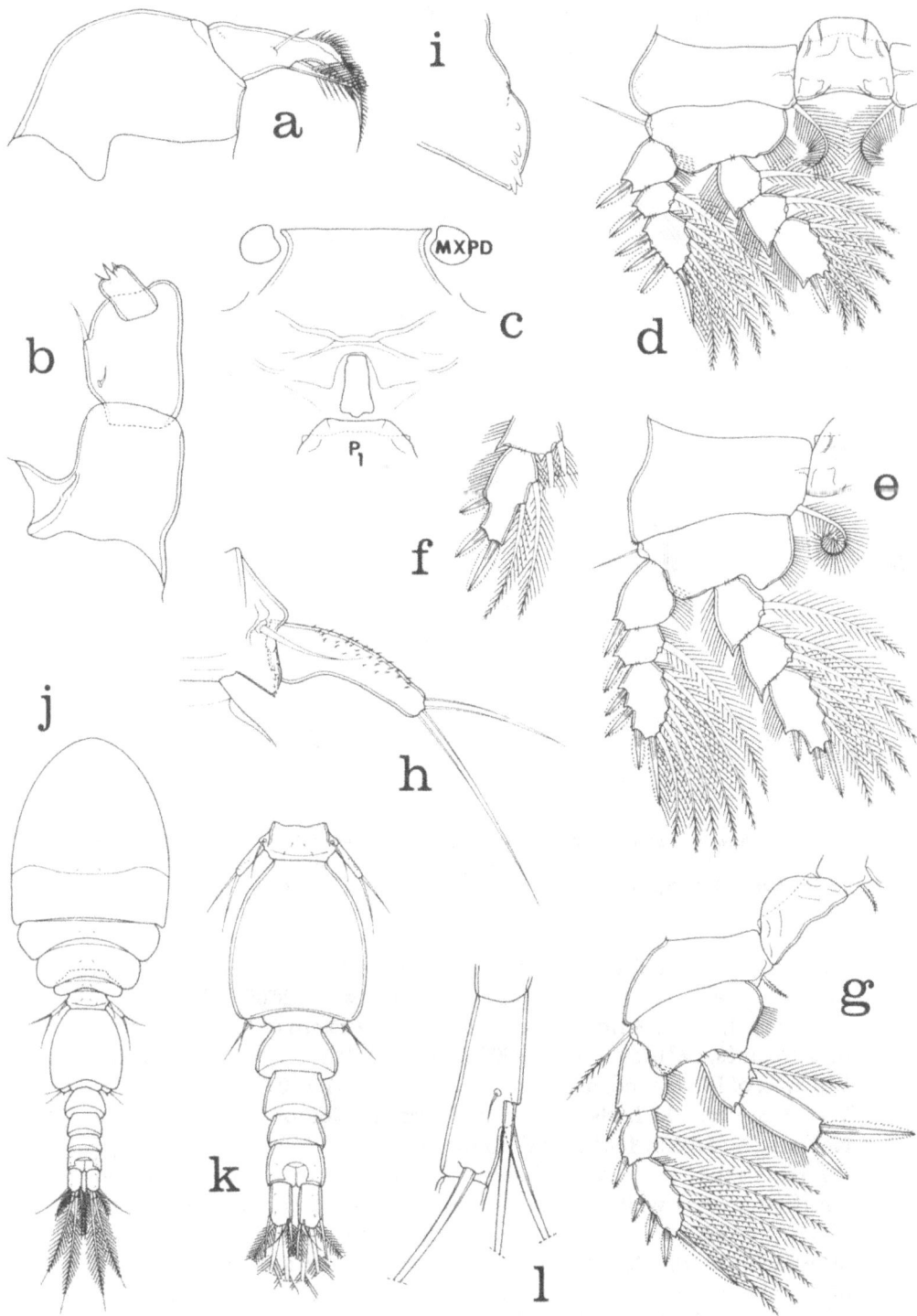

FIG. 11. *Doridicola caelatus* sp. n. Female: a, second maxilla, posterior (F); b, maxilliped, inner (F); c, area between maxillipeds and first pair of legs, ventral (E); d, leg 1 and intercoxal plate, anterior (E); e, leg 2, anterior (E); f, third segment of endopod of leg 3, anterior (E); g, leg 4 and intercoxal plate, anterior (E); h, leg 5, dorsal (D); i, detail of wing on segment of leg 5, dorsal (G). Male: j, dorsal (A); k, urosome, dorsal (B); l, fourth segment of first antenna, ventral (C).

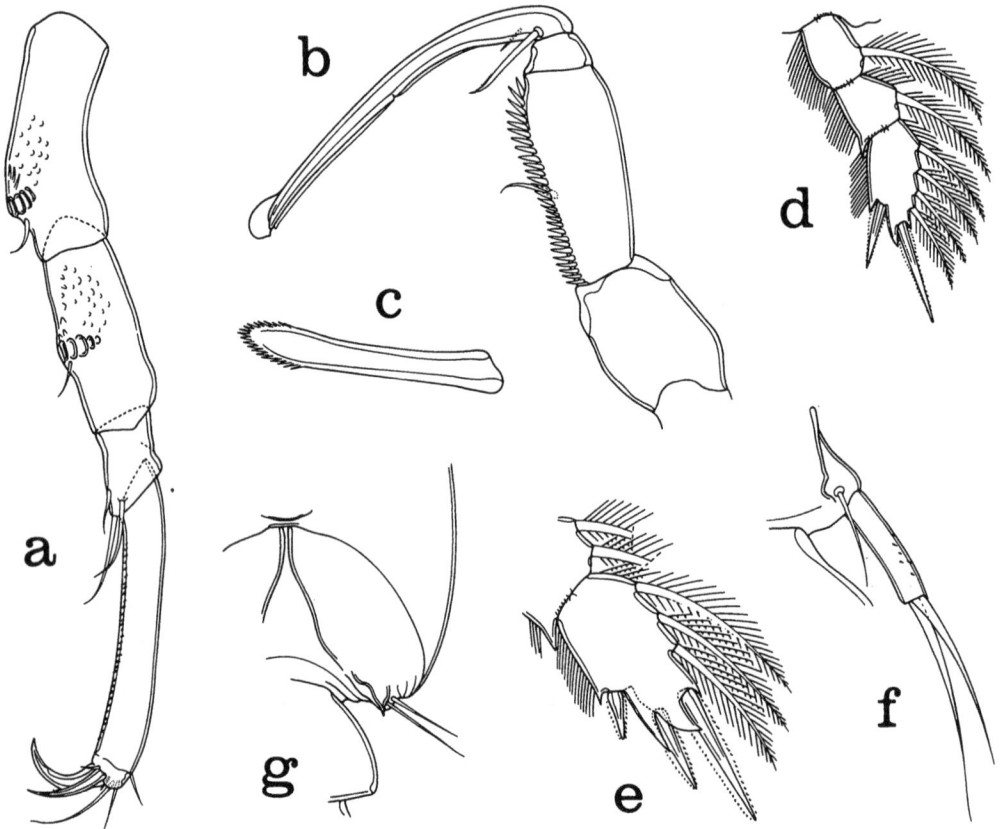

FIG. 12. *Doridicola caelatus* sp. n., male: a, second antenna, anterior (F); b, maxilliped, postero-inner (D); c, spine on second segment of maxilliped, antero-outer (H); d, endopod of leg 1, anterior (D); e, third segment of endopod of leg 2, anterior (F); f, leg, 5, dorsal (F); g, leg 6, ventral (D).

Other specimens (all from *Entacmaea quadricolor*).—1 ♂ from three hosts, in 1 m, western side of Isle Maître, near Nouméa, 22°20′05″S, 166°24′05″E, 11 June 1971; 3 ♀♀, 1 ♂ from three hosts, in 1 m, western end of Isle Mando, near Nouméa, 22°18′59″S, 166°09′30″E, 3 July 1971; 1 ♀, 4 ♂♂ from 12 hosts, 1 m, west of Isle Mando, near Nouméa, 22°18′59″S, 166°09′30″E, 5 July 1971, 6 ♀♀, 9 ♂♂ from 15 hosts, in tide pool, eastern side of Isle Maître, near Nouméa, 22°20′35″S, 166°25′10″E, 16 July 1971.

Female.—Body (Fig. 10a) moderately slender. Length 1.81 mm (1.70–1.91 mm) and greatest width 0.69 mm (0.66–0.73 mm), based on 10 specimens. Segment of leg 1 separated from cephalosome by dorsal furrow. Epimera of segments bearing legs 1–4 rounded posteriorly. Ratio of length to width of prosome 1.52:1. Ratio of length of prosome to that of urosome 1.29:1.

Segment of leg 5 (Fig. 10b) 104 × 224 μm. Genital segment elongate, 320 μm long, 213 μm wide in its anterior third and 161 μm wide in its posterior two-thirds where sides are nearly parallel. Genital areas situated dorsolaterally in widened anterior third of segment. Each area (Fig. 10c) with two small naked setae about 7 μm long. Three postgenital segments from anterior to posterior 153 × 143, 86 × 122, and 122 × 117 μm. Posteroventral margin of anal segment with row of extremely minute spinules on each side.

Caudal ramus (Fig. 10d) moderately elongate, 109 × 50 μm, ratio 2.18:1. Outer lateral seta 275 μm and naked. Dorsal seta approximately 80 μm and haired. Outermost terminal seta 145 μm and innermost terminal seta 242 μm, both with lateral hairs. Two long median terminal setae 363 μm (outer) and 440 μm (inner), both with long lateral spinules and both inserted between dorsal flange (smooth) and ventral flange (with marginal row of extremely minute spinules).

Body surface with relatively few hairs (sensilla) except on rostral area where more numerous long hairs occur (Fig. 10a).

Egg sac (Fig. 10e) elongate, 650 × 264 μm, containing approximately 27 eggs of irregular outline from 100–138 μm in diameter.

Rostrum (Fig. 10f) with broad round posteroventral edge, its surface bearing many long hairs and refractile points.

First antenna (Fig. 10g) 537 μm long. Lengths of seven segments: 39 (86 μm along anterior margin), 133, 38, 83, 82, 68, and 47 μm respectively. Formula for armature as in *Doridicola cylichnophorus*. All setae naked.

Second antenna (Fig. 10h) 4-segmented and 420 μm long. Formula for armature as in *Doridicola cylichnophorus*. Fourth segment long and slender, 161 μm along outer side, 120 μm along inner side, and 24 μm wide. Two terminal claws slightly unequal, stouter claw 52 μm, more slender claw 44 μm with a few small teeth (spinules) on concave margin (Fig. 10i). All four segments with minute inner spinules. All setae naked.

Labrum (Fig. 10j) with two rounded posteroventral lobes. Mandible (Fig. 10k), paragnath (Fig. 10i), first maxilla (Fig. 10l) and second maxilla (Fig. 11a) similar in general respects to species of *Doridicola* described above. Maxilliped (Fig. 11b) with first two segments stout and third segment small. Armature as in previously described species.

Ventral area between maxillipeds and first pair of legs (Fig. 11c) slightly protuberant.

Legs 1–4 (Fig. 11d–g) segmented and armed as in previously described species of *Doridicola*. Outer seta on basis of leg 3 with lateral hairs. Leg 4 (Fig. 11g) with exopod 204 μm long. First endopod segment 57 × 52 μm, its inner distal plumose seta 109 μm. Second segment 96 × 44 μm, its two terminal fringed spines 52 μm (outer) and 100 μm (inner). Inner coxal seta 30 μm with short lateral spinules.

Leg 5 (Fig. 11h) with moderately elongate free segment 125 μm long, 41 μm wide at inner proximal expansion and 30 μm wide distally. Two terminal smooth setae 170 μm (inner) and 117 μm (outer). Dorsal seta approximately 73 μm and smooth. Posterolateral area of segment bearing leg 5 projected and bearing several small spines (Fig. 11i).

Leg 6 represented by two setae on genital area (Fig. 10c).

Color in living specimens in transmitted light opaque gray, eye red, egg sacs bright lavender.

Male.—Body (Fig. 11j) resembling that of female in general form. Length 1.35 mm (1.32–1.42 mm) and greatest width 0.44 mm (0.42–0.47 mm), based on 10 specimens. Ratio of length to width of prosome 1.56:1. Ratio of length of prosome to that of urosome 1.28:1.

Segment of leg 5 (Fig. 11k) 57 × 125 μm. Genital segment 247 × 213 μm, longer than wide. Four postgenital segments from anterior to posterior 73 × 117, 68 × 101, 42 × 88, and 60 × 84 μm.

Caudal ramus (Fig. 11k) 68 × 36 μm, ratio 1.89:1, smaller than in female.

Rostrum as in female. First antenna apparently as in female, without added aesthetes, though one male first antenna showed small additional element on fourth segment (Fig. 11l).

Second antenna (Fig. 12a) 330 μm long, segmented and armed as in female but showing sexual dimorphism. First and second segments with embossed saucerlike rings and scattered scalelike ridges. Fourth segment with short inner spinules.

Labrum, mandible, paragnath, first maxilla, and second maxilla as in female. Maxilliped (Fig. 12b) segmented as in previously described species of *Doridicola*. First segment unarmed. Second segment with naked seta and modified spine with spinulose tip (Fig. 12c) and bearing inner row of spines. Small third segment unornamented. Claw 254 μm along its axis including terminal lamella, bearing two unequal proximal setae.

Ventral area between maxillipeds and first pair of legs as in female.

Legs 1–4 segmented and armed as in female, except for third endopod segment of leg 1 with formula I, I, 4 (Fig. 12d). Sexual dimorphism also seen in third endopod segment of leg 2 (Fig. 12e) with its large spiniform processes. Legs 3 and 4 like those of female.

Leg 5 (Fig. 12f) with free segment 55 × 13 μm, without inner proximal expansion.

Leg 6 (Fig. 12g) a posteroventral flap on genital segment bearing two naked setae 47 μm and 55 μm.

Spermatophore unknown.

Color in living specimens as in female.

Etymology.—The specific name *caelatus*, from Latin *caelare* meaning to emboss, refers to the embossed appearance of the saucerlike rings on the second antenna of the male.

Remarks.—*Doridicola caelatus* may be distinguished from most species in the genus by a combination of three characters in the female: the ratio of the caudal ramus, the form of the genital segment, and the nature of the free segment of leg 5 together with the adjacent spinous lobe. Only *Doridicola actiniae* (Della Valle, 1880a) appears to be close to the new species. In contrast to *D. caelatus*, however, this Mediterranean species from *Actinia equina* and *Anemonia sulcata* has four setae on the first maxilla, the fourth segment of the second antenna has a ratio of about 4:1, the first two segments of the second antenna of the male apparently lack embossed rings, and the third endopod segment of legs 1 and 2 in the male have prominent spiniform processes with granular surfaces (*fide* redescription by Stock, 1960).

The saucerlike rings on the first two segments of the male second antenna are diagnostic of *D. caelatus*. Such ornamentation occurs in no other species of *Doridicola* as far as known. As noted above, males are unknown in four species of this genus, and in two species the description of the male is so incomplete that sexual dimorphism in the second antenna cannot be determined.

Doridicola hispidulus sp. n
Figs. 13a–k, 14a–h, 15a–i

Type material.—3 ♀♀, 7 ♂♂ from three individuals of the sea anemone *Entacmaea quadricolor* (Reuppell and Leuckart), in 1 m, western side of Isle Maître, near Nouméa, New Caledonia, 22°20′05″S, 166°24′05″E, 11 June 1971. Holotype ♀ allotype, and 7 paratypes (2 ♀♀, 5 ♂♂) deposited in the National Museum of Natural History, Smithsonian Institution, Washington, D.C.; the remaining paratypes (dissected) in the collection of the author.

Other specimens (all from *Entacmaea quadricolor*).—4 ♀♀ from 3 hosts, in 1 m, western end of Isle Mando, near Nouméa, 22°18′59″S, 166°09′30″E, 3 July 1971; 1 ♂ from 5 hosts, in 1 m, western end of Isle Mando, near Nouméa, 22°18′59″S, 166°09′30″E, 3 July 1971; 2 ♀♀ from 12 hosts, in 1 m, west of Isle Mando, near Nouméa, 22°18′59″S, 166°09′30″E, 5 July 1971; 2 ♀♀, 2 ♂♂ from 5 hosts, in tide pool, eastern end of Isle Maître, near Nouméa, 22°20′35″S, 166°25′10″E, 16 July 1971.

Female.—Body (Fig. 13a) with broad prosome. Length 1.62 mm (1.54–1.70 mm) and greatest width 0.82 mm (0.80–0.89 mm), based on 10 specimens. Segment of leg 1 separated from cephalosome by dorsal furrow. Epimera of segments bearing legs 1–4 rounded posteriorly. Ratio of length to width of prosome 1.25:1. Ratio of length of prosome to that of urosome 1.78:1.

Segment of leg 5 (Fig. 13b) 78 × 211 μm. Genital segment 208 × 231 μm, wider than long, expanded in its anterior two-thirds. Genital areas located dorsolaterally in anterior half of segment. Each area (Fig. 13c) with two small naked setae 9 μm and 13 μm, two minute spiniform processes, and a field of small spines. Three postgenital segments from anterior to posterior 86 × 117, 65 × 112, and 88 × 117 μm. Posteroventral margin of anal segment with row of very small spinules on each side.

Caudal ramus (Fig. 13d) moderately elongate, 99 × 50 μm, ratio 1.98:1. Outer lateral seta 300 μm and naked. Dorsal seta about 78 μm with short hairs. Outermost terminal seta 130 μm with hairs along outer side. Innermost terminal seta 360 μm with rather widely spaced lateral spinules. Two long median terminal setae 539 μm (outer) and 560 μm (inner), both with coarse widely spaced lateral spinules and inserted between dorsal flange (smooth) and ventral flange (with marginal row of very small spinules).

Body surface with small number of hair (sensilla) as in Figure 13a.

Egg sac (Fig. 13e) elongate, 715 × 253 μm, containing approximately 30 somewhat irregularly shaped eggs 100–130 μm in diameter.

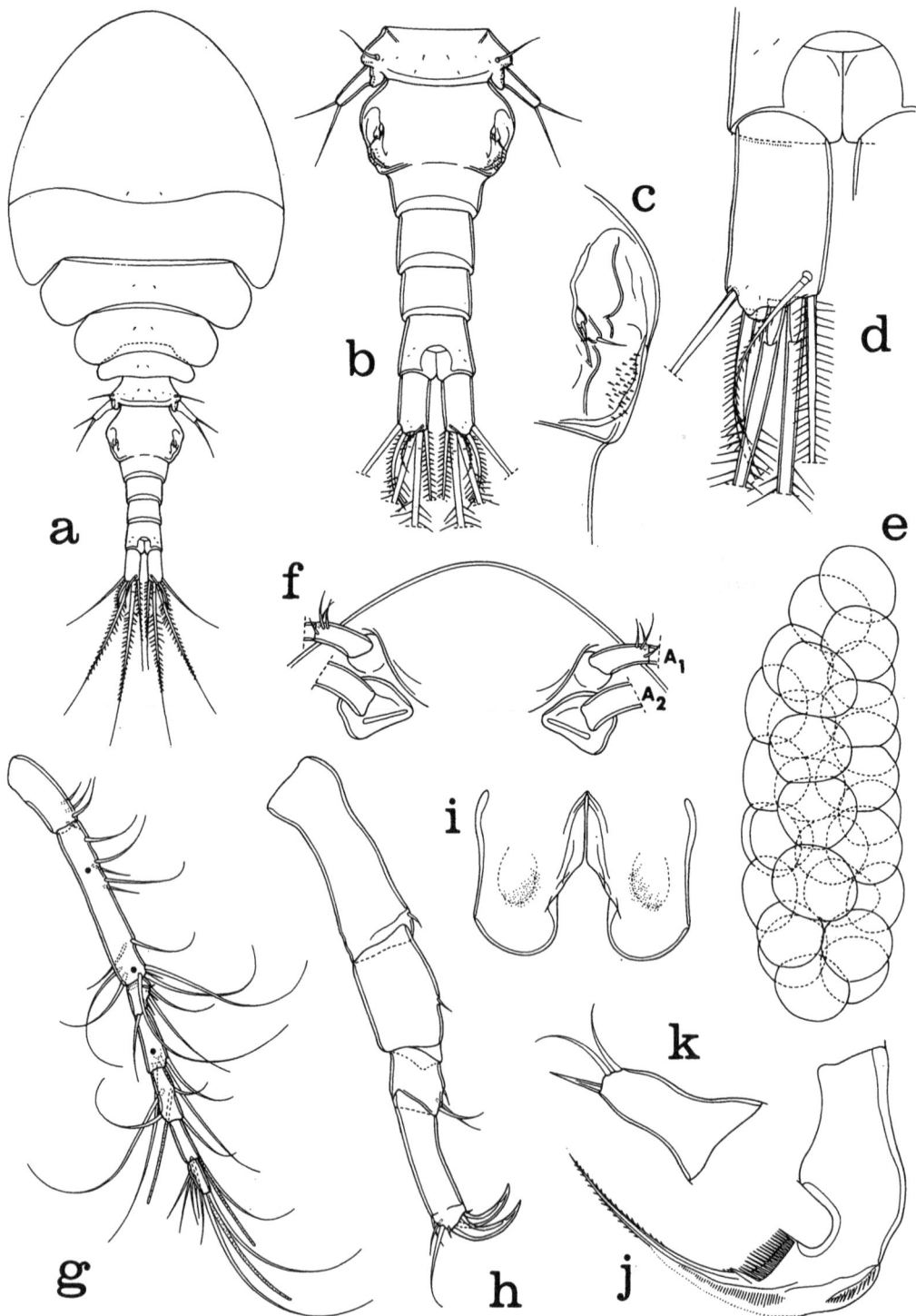

FIG. 13. *Doridicola hispidulus* sp. n., female: a, dorsal (A); b, urosome, dorsal (B); c, genital area, dorsal (F); d, caudal ramus, dorsal (F); e, egg sac, ventral (B); f, rostrum, ventral (B); g, first antenna, with dots indicating positions of aesthetes added in male, dorsal (E); h, second antenna, anterior (D); i, labrum, with paragnaths indicated by broken lines, ventral (F); j, mandible, posterior (C); k, first maxilla, posterior (C).

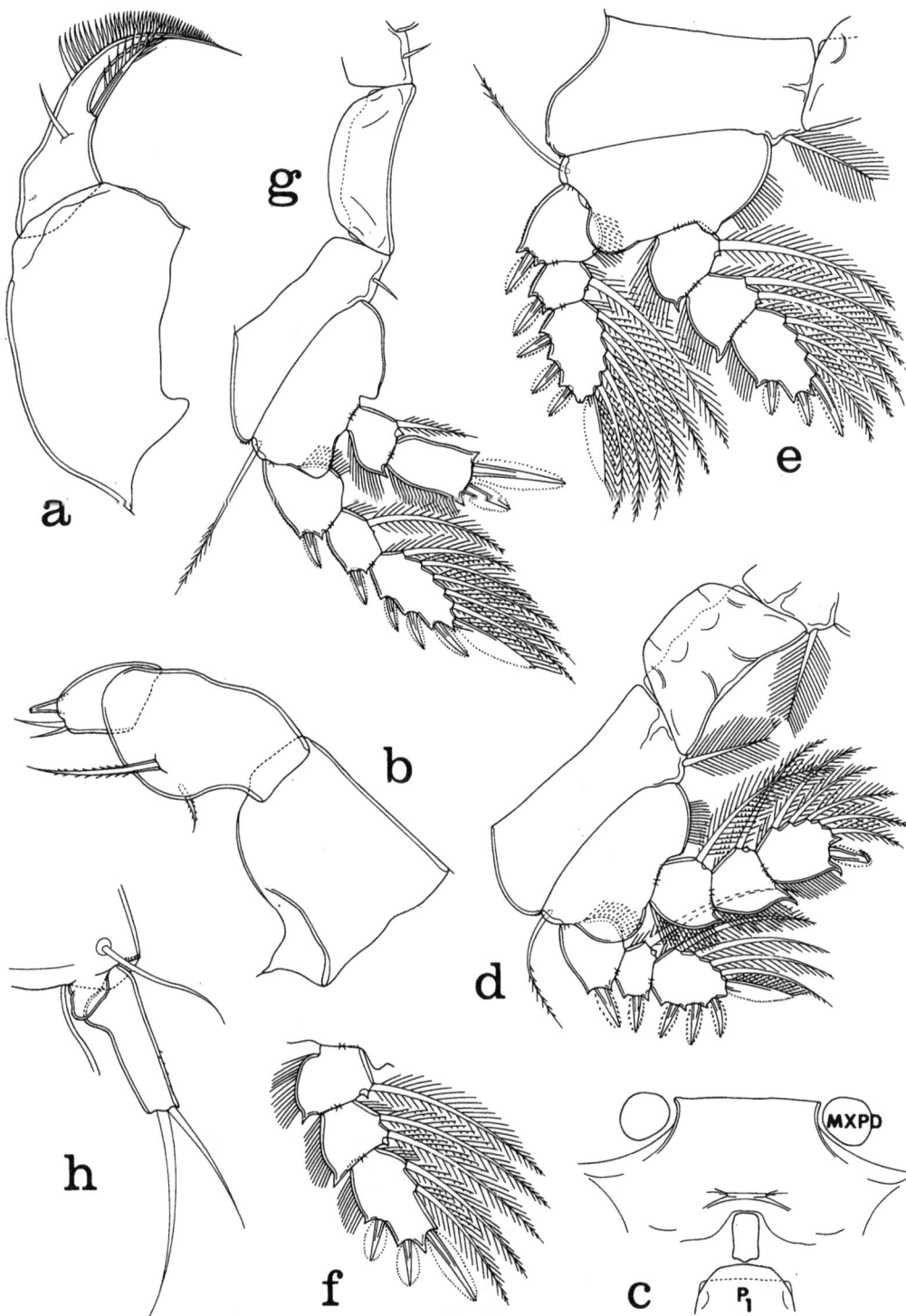

FIG. 14. *Doridicola hispidulus* sp. n., female: a, second maxilla, posterior (C); b, maxilliped, postero-inner (C); c, area between maxillipeds and first pair of legs, ventral (E); d, leg 1 and intercoxal plate, anterior (D); e, leg 2, anterior (D); f, endopod of leg 3, anterior (D); g, leg 4 and intercoxal plate, anterior (D); h, leg 5, dorsal (F).

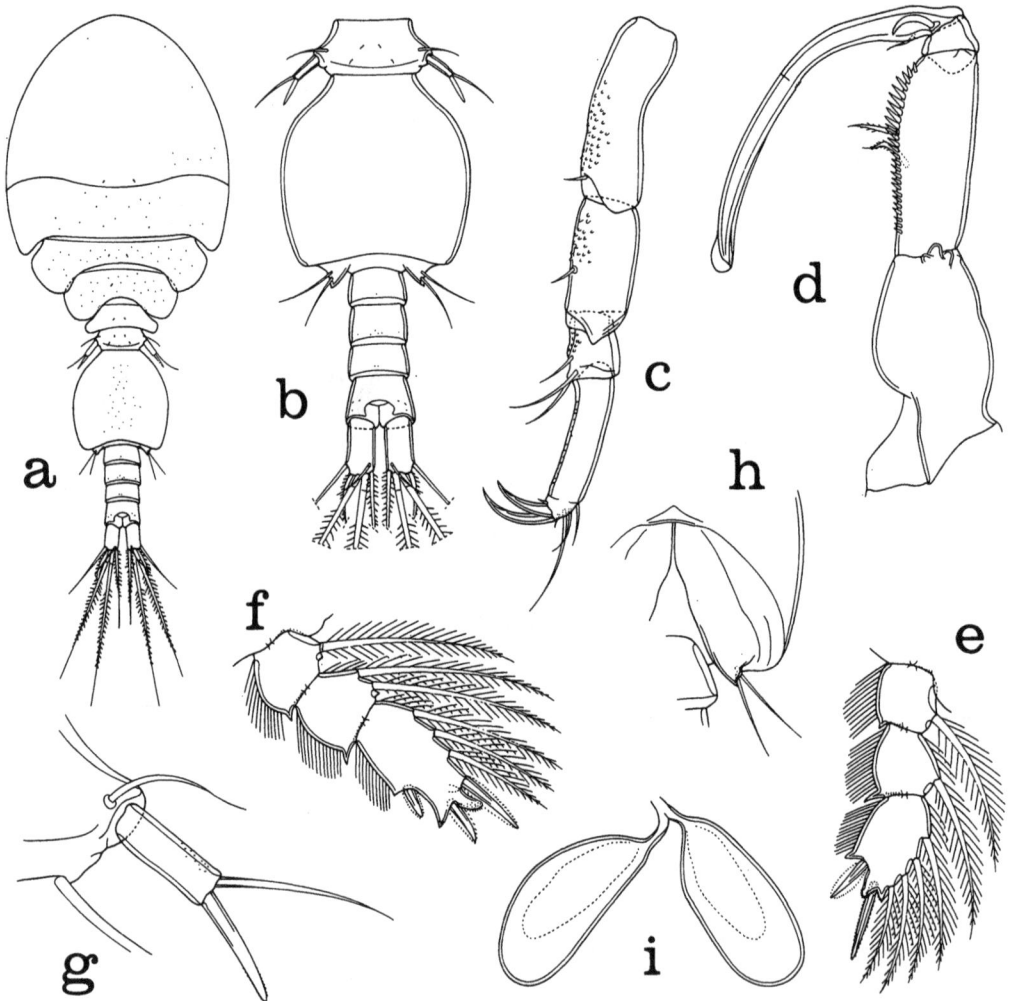

FIG. 15. *Doridicola hispidulus* sp. n., male: a, dorsal (A); b, urosome, dorsal (B); c, second antenna, posterior (D); d, maxilliped, postero-inner (D); e, endopod of leg 1, anterior (D); f, endopod of leg 2, anterior (D); g, leg 5, dorsal (C); h, leg 6, ventral (E); i, spermatophores, attached to female, dorsal (B).

Rostrum (Fig. 13f) without well-defined posteroventral border.

First antenna (Fig. 13g) 496 μm long. Lengths of seven segments: 44 (88 μm along anterior margin), 174, 34, 62, 55, 44, and 39 μm respectively. Formula for armature as in preceding species of *Doridicola*. All setae naked.

Second antenna (Fig. 13h) 355 μm long and 4-segmented, with first segment unusually long (117 μm) in comparison with second segment (70 μm). Formula for armature as in preceding species of *Doridicola*. Fourth segment elongate, 104 μm along outer side, 70 μm along inner side, and 23.5 μm wide. Two terminal claws nearly equal, 49 μm and 52 μm. Fine ornamentation lacking and all setae smooth.

Labrum (Fig. 13i) with two elongate round posteroventral lobes. Mandible (Fig. 13j) resembling in general form that of species of *Doridicola* described above, but row of spinules on concave side of base with spinules on ends of row a little stronger and more spinelike than others. Paragnath (Fig 15i) a small hairy lobe. First maxilla (Fig. 13k) with three smooth terminal elements, two setiform and one spiniform. Second maxilla (Fig. 14a) similar to that in preceding species of *Doridicola;* lash with long slender graduated spines. Maxilliped (Fig. 14b) resembling in general aspect that of preceding species of *Doridicola;* two setae on second segment with short lateral spinules.

Ventral area between maxillipeds and first pair of legs (Fig. 14c) not protuberant.

Legs 1–4 (Fig. 14d–g) segmented and armed as in genus *Doridicola*. Leg 1 with tip of spine on third endopod segment recurved posteriorly (Fig. 14d). Leg 2 with three spines on third endopod segment having tips recurved posteriorly. Leg 4 (Fig. 14g) with exopod 174 μm long. Endopod unusually short, about 100 μm. First segment 39 × 40 μm, its inner plumose seta 57 μm. Second segment 60 × 34 μm, its two terminal spines 36 μm (outer) and 69 μm (inner). Inner coxal seta 19 μm and naked.

Leg 5 (Fig. 14h) with free segment 75 μm long, 31 μm wide at proximal inner rounded expansion, and 18 μm wide distally. Two terminal setae 112 μm (inner) and 70 μm (outer). Dorsal seta about 78 μm. All setae smooth. Free segment ornamented with few minute spines on distal outer surface. Dorsally near insertion of free segment an outer row of small spines and an inner irregularly shaped unornamented lobe.

Leg 6 represented by two setae on genital area (Fig. 13c).

Color in living specimens in transmitted light opaque gray, eye red, egg sacs gray.

Male.—Body (Fig. 15a) with relatively broad prosome as in female. Length 1.33 mm (1.21–1.41 mm) and greatest width 0.56 mm (0.53–0.63 mm), based on 10 specimens. Ratio of length to width of prosome 1.30:1. Ratio of length of prosome to that of urosome 1.39:1.

Segment of leg 5 (Fig. 15b) 70 × 164 μm. Genital segment 265 μm long including leg 6 (247 μm without leg 6) and 268 μm wide. Four postgenital segments from anterior to posterior 52 × 94, 52 × 91, 39 × 91, and 65 × 99 μm.

Caudal ramus (Fig. 15b) 78 × 44 μm, ratio 1.77:1, smaller than in female.

Rostrum as in female. First antenna similar to that of female but three aesthetes added (at points indicated by dots in Figure 13g) so that formula is: 4, 13 + 2 aesthetes, 6, 3 + 1 aesthete, 4 + 1 aesthete, 2 + 1 aesthete, and 7 + 1 aesthete.

Second antenna (Fig. 15c) 330 μm long, segmented and armed as in female but sexually dimorphic in having very small spines on inner surface of all four segments.

Labrum, mandible, paragnath, first maxilla, and second maxilla like those of female. Maxilliped (Fig. 15d) segmented as in previously described species of *Doridicola*. First segment unarmed but showing distally three small scle-rotized lobes. Second segment with inner row of spines and two delicately barbed setae. Small third segment unarmed. Claw 230 μm along its axis including terminal lamella, with concave margin at level of weak subdivision minutely crenate; two proximal elements more nearly equal in length than in most *Doridicola*, larger element stout with minutely barbed tip.

Ventral area between maxillipeds and first pair of legs as in female.

Legs 1–4 segmented and armed as in female, but third endopod segment of leg 1 with formula I, I, 4 (Fig. 15e). Sexual dimorphism also apparent in third endopod segment of leg 2 (Fig. 15f) with its enlarged terminal processes. Legs 3 and 4 like those of female.

Leg 5 (Fig. 15g) with rectangular free segment 42 × 13 μm, without proximal inner expansion. Two terminal elements, inner spiniform (39 μm) with minutely barbed tip, outer setiform (58 μm) and smooth. Dorsal seta approximately 50 μm. Row of spinules and lobes seen in female near insertion of free segment here absent.

Leg 6 (Fig. 15h) a posteroventral flap on genital segment bearing two naked setae about 92 μm long.

Spermatophore (Fig. 15i) about 255 × 114 μm without neck.

Color in living specimens as in female.

Etymology.—The specific name *hispidulus*, from Latin *hispidus* meaning bristly or prickly and the diminutive suffix *-ulus*, alludes to the patch of small spines on the genital area of the female.

Remarks.—*Doridicola hispidulus* and *Doridicola caelatus* share in common one character not found as far as known in other species in the genus, namely the dorsal lobe at the inner side of the insertion of the free segment of leg 5. In contrast to *D. caelatus*, however, this lobe in *D. hispidulus* is unornamented.

The new species differs from all its congeners in having a patch of spines on the genital area of the female.

Doridicola penicillatus sp. n.
Figs. 16a–k, 17a–j, 18a–f

Type material.—1 ♀, 1 ♂ from five specimens of the sea anemone *Entacmaea quadricolor* (Rueppell and Leuckart), in 1 m, western end of Isle Mando, near Nouméa, New Caledonia, 22°18′59″S, 166°09′30″E, 3 July 1971. Holotype ♀ and allotype deposited in the National Museum of Natural History, Smithsonian Institution, Washington, D.C.

Other specimens (all from *Entacmaea quadricolor*).—1 ♂ (dissected) from 3 hosts, in 1 m, western end of Isle Mando, 3 July 1971; 1 ♀ (dissected) from 12 hosts, in 1 m, west of Isle Mando, 5 July 1971; 1 ♀ from 5 hosts, in tide pool, eastern end of Isle Maître, 22°20′35″S, 166°25′10″E, 16 July 1971.

Female.—Body (Fig. 16a) with broad prosome. Length 1.67 mm (1.52–1.77 mm) and greatest width 0.77 mm (0.73–0.80 mm), based on 3 specimens. Segment of leg 1 clearly separated from cephalosome by dorsal furrow. Epimera of segments of legs 1–4 rounded posteriorly. Ratio of length to

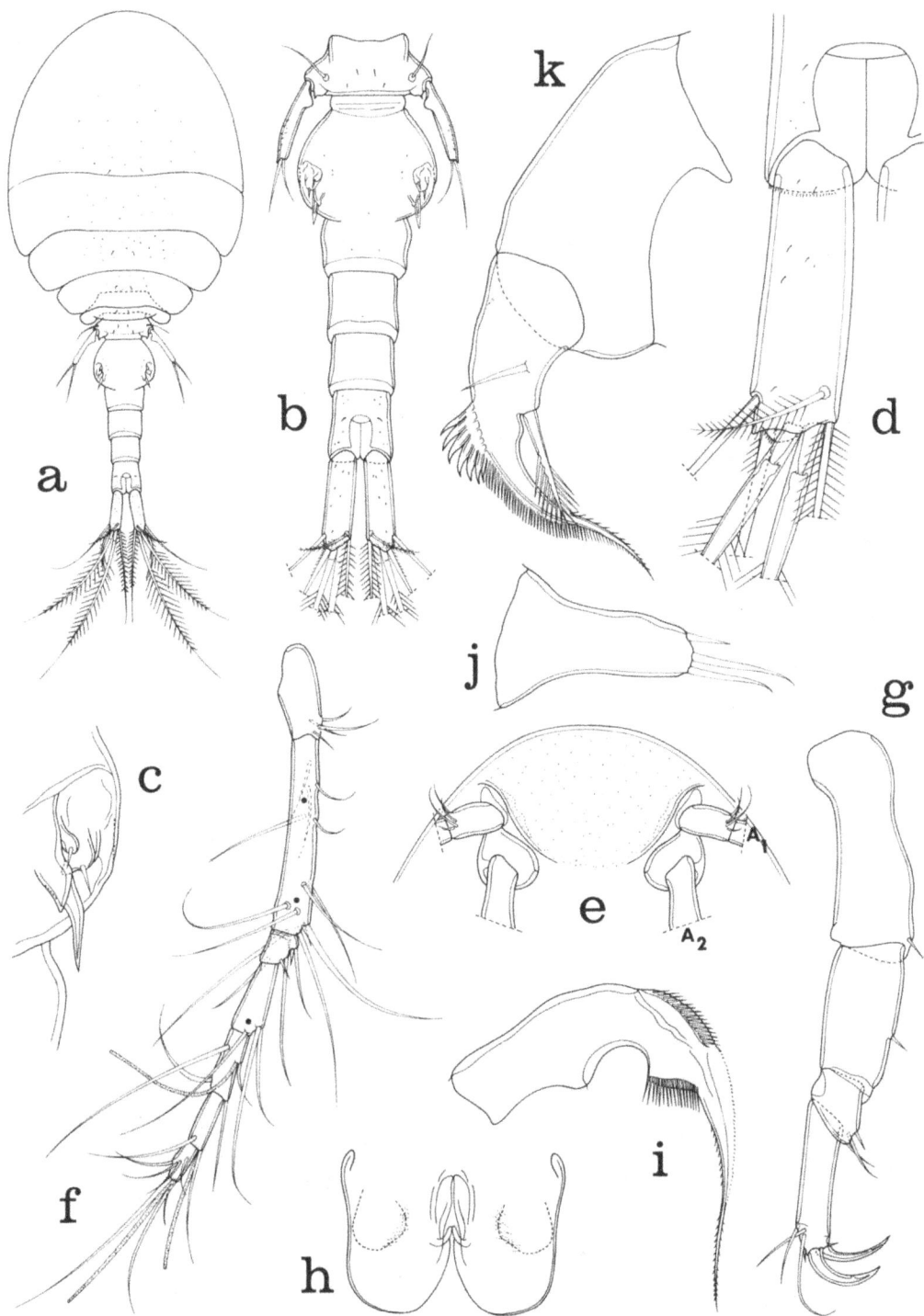

FIG. 16. *Doridicola penicillatus* sp. n., female: a, dorsal (A); b, urosome, dorsal (B); c, genital area, dorsal (F); d, caudal ramus, dorsal (F); e, rostrum, ventral (B); f, first antenna, with dots indicating positions of aesthetes in male, ventral (E); g, second antenna, anterior (D); h, labrum, with paragnaths indicated by broken lines, ventral (F); i, mandible, posterior (C); j, first maxilla, posterior (G); k, second maxilla, posterior (C).

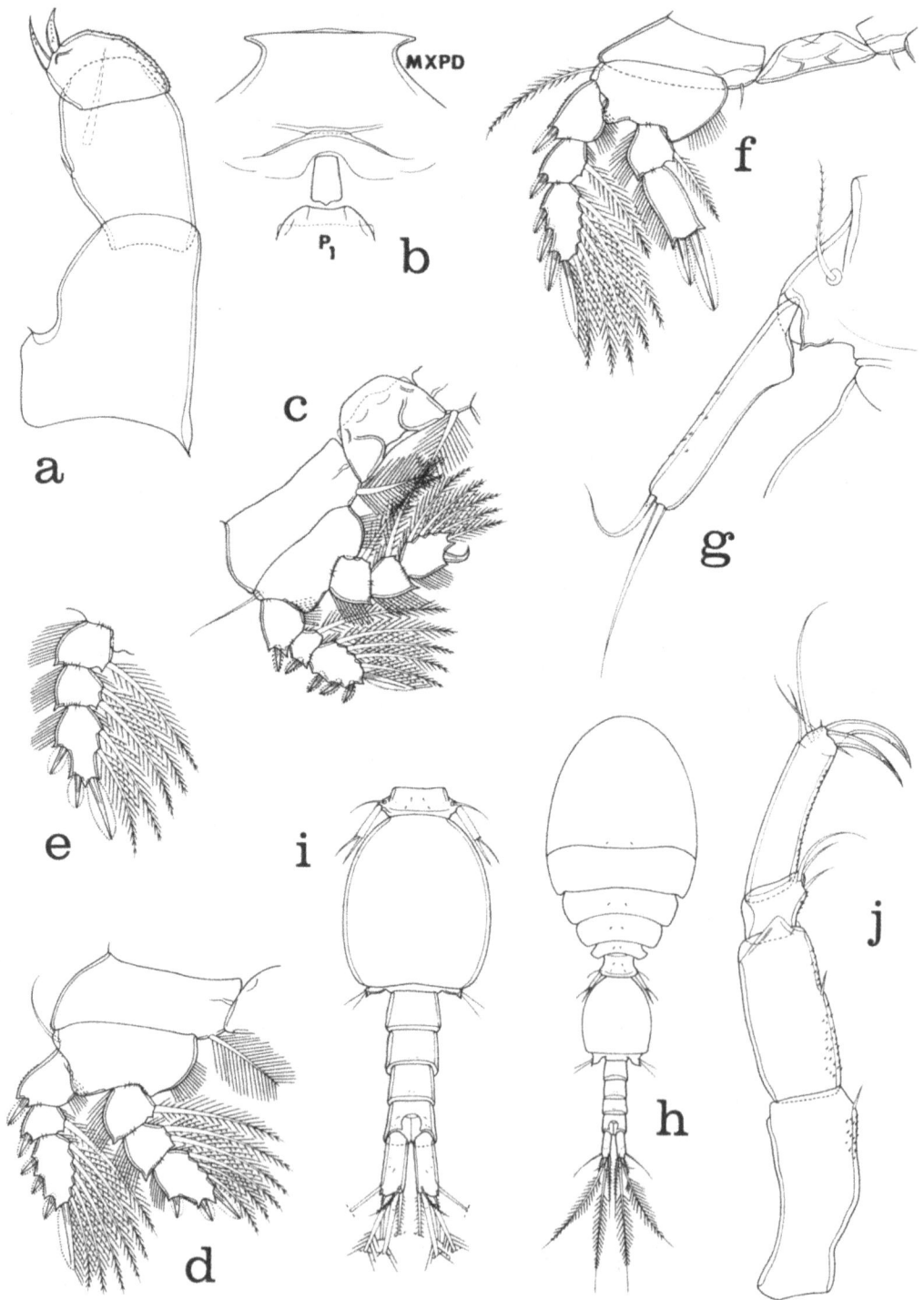

FIG. 17. *Doridicola penicillatus* sp. n., Female: a, maxilliped, antero-outer (C); b, area between maxillipeds and first pair of legs, ventral (E); c, leg 1 and intercoxal plate, anterior (E); d, leg 2, anterior (E); e, endopod of leg 3, anterior (E); f, leg 4 and intercoxal plate, anterior (E); g, leg 5, dorsal (F). Male: h, dorsal (A); i, urosome, dorsal (B); j, second antenna, posterior (F).

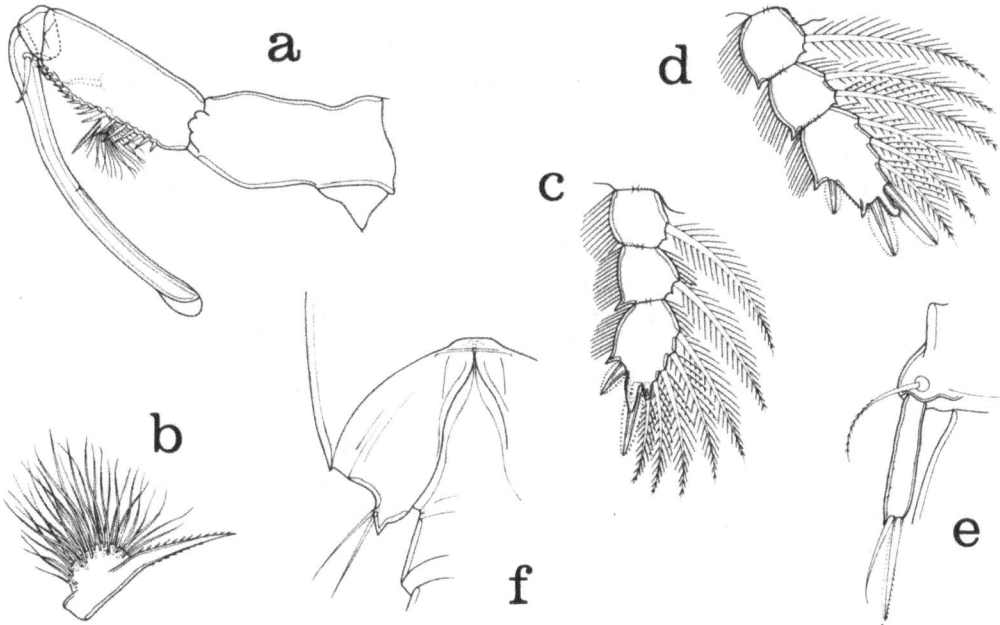

FIG. 18. *Doridicola penicillatus* sp. n., male: a, maxilliped, postero-inner (D); b, seta on second segment of maxilliped, antero-outer (G); c, endopod of leg 1, anterior (D); d, endopod of leg 2, anterior (D); e, leg 5, dorsal (F); f, leg 6, ventral (D).

width of prosome 1.26:1. Ratio of length of prosome to that of urosome 1.39:1.

Segment of leg 5 (Fig. 16b) 96 × 205 μm. Genital segment 234 μm long, expanded with rounded lateral margins in anterior two-thirds (width 210 μm) and with subparallel lateral margins in posterior third (width 134 μm). Genital areas located dorsolaterally near middle of segment. Each area (Fig. 16c) bearing two small naked setae approximately 11 μm long and small spiniform process. Medially and posteriorly to genital area a subtriangular spiniform process and a long prominent slender spiniform process reaching well beyond edge of genital segment. Three postgenital segments from anterior to posterior 99 × 109, 78 × 100, and 104 × 101 μm. Posteroventral margin of anal segment with row of minute spinules on each side.

Caudal ramus (Fig. 16d) elongate, 151 × 42 μm, ratio 3.60:1. Outer lateral seta 264 μm and naked. Dorsal seta about 78 μm and haired. Outermost terminal seta 160 μm, with short hairs along proximal outer side. Innermost terminal seta 319 μm with lateral hairlike spinules. Two long median terminal setae 605 μm (outer) and 715 μm (inner), both with widely spaced lateral spinules and inserted between dorsal flange (smooth) and ventral flange (with marginal row of slender spinules).

Body surface with few hairs (sensilla) and refractile points as in Figure 16a.

Egg sac unknown.

Rostrum (Fig. 16e) broad, without well-defined posteroventral border.

First antenna (Fig. 16f) 570 μm long. Lengths of seven segments: 49 (96 μm along anterior margin), 195, 34, 75, 68, 57, and 45 μm respectively. Formula for armature as in preceding species of *Doridicola*. All setae naked.

Second antenna (Fig. 16g) 385 μm long including claws and 4-segmented. Formula for armature as in preceding species of *Doridicola*. Fourth segment elongate, 107 μm along outer side, 70 μm along inner side, and 24 μm wide. Two smooth terminal claws nearly equal in length, 53 μm and 57 μm. All setae smooth and fine ornamentation of segments lacking.

Labrum (Fig. 16h) with two rounded posteroventral lobes. Mandible (Fig. 16i) with rather long slender spines on scalelike area of base. Paragnath (Fig. 16h) a small hairy lobe. First maxilla (Fig. 16j) with three naked terminal setae. Second maxilla (Fig. 16k) with lash bearing several prominent stout pointed teeth (resembling cock's comb), followed by long series of much reduced slender spinules. Maxilliped (Fig. 17a) with two smooth very unequal setae on second segment; third segment bearing two terminal elements with few spinules and having on its outer surface numerous small spines.

Ventral area between maxillipeds and first pair of legs (Fig. 17b) not protuberant.

Legs 1–4 (Fig. 17c–f) segmented and armed as in species of *Doridicola* described above. Leg 1 with spine on third endopod segment falciform and smooth (Fig. 17c). Leg 4 (Fig. 17f) with exopod 180 μm long. Endopod relatively short, first segment 49 × 48 μm and its inner plumose seta 99 μm, second segment 81 × 41 μm, its two terminal fringed spines 45 μm (outer) and 84 μm (inner). Inner coxal seta 26 μm and naked. Outer seta on basis haired.

Leg 5 (Fig. 17g) with elongate slender free segment 122 μm long, 28 μm wide at proximal inner expansion and 21 μm wide distally. Ratio 4.36:1 (width at expansion) or 5.81:1 (width distally). Two terminal smooth setae 57 μm and 90 μm. Dorsal seta 60 μm and haired. Free segment with few very small spines on outer dorsal surface. Dorsally near insertion of free segment an inner mammilliform lobe with small acuminate tip.

Leg 6 represented by two setae on genital area (Fig. 16c).

Color in living specimens in transmitted light opaque gray, eye red.

Male.—Body (Fig. 17h) with prosome much more slender than in female. Length 1.33 mm (1.32–1.33 mm) and greatest width 0.49 mm (0.48–0.52 mm), based on 2 specimens. Ratio of length to width of prosome 1.54:1. Ratio of length of prosome to that of urosome 1.19:1.

Segment of leg 5 (Fig. 17i) 47 × 130 μm. Genital segment 273 μm long (including leg 6) and 229 μm wide, a little longer than wide. Four postgenital segments from anterior to posterior 52 × 86, 57 × 78, 47 × 70, and 65 × 76 μm.

Caudal ramus (Fig. 17i) 104 × 35 μm, ratio 2.97:1, smaller than in female.

Rostrum as in female. First antenna resembling that of female but three aesthetes added (at points indicated by dots in Figure 16f) so that formula is same as in *Doridicola hispidulus* above.

Second antenna (Fig. 17j) 286 μm long, resembling that of female but sexually dimorphic in having small spines on inner surface of all four segments.

Labrum, mandible, paragnath, first maxilla, and second maxilla like those of female. Maxilliped (Fig. 18a) segmented as in previously described species of *Doridicola*. First segment unarmed but showing distally three small lobes as in *Doridicola hispidulus*. Second segment with inner row of spines and two setae, one smooth, other elaborately modified (Fig. 18b), with unilaterally swollen base having long slender setules collectively with brushlike appearance. Small third segment unarmed. Claw 218 μm along its axis including terminal lamella, with two unequal proximal setae, larger seta smooth.

Ventral area between maxillipeds and first pair of legs as in female.

Legs 1–4 segmented and armed as in female, but third endopod segment of leg 1 with formula I, I, 4 (Fig. 18c), outer terminal leaflike process with extremely small spines, inner terminal spiniform process with its tip turned strongly anteriorly. Third endopod segment of leg 2 with modified terminal process (Fig. 18d).

Leg 5 (Fig. 18e) with slender elongate free segment 57 × 13 μm, without proximal inner expansion, ratio 4.38:1. Two terminal segments, inner element spiniform (47 μm) and delicately barbed, outer element setiform (42 μm) and smooth. Dorsal seta 52 μm with short hairs. Mammilliform lobe seen in female near free segment here absent.

Leg 6 (Fig. 18f) a posteroventral flap on genital segment bearing two smooth setae approximately 65 μm long.

Spermatophore not seen.

Color in living specimens as in female.

Etymology.—The specific name *penicillatus*, from Latin *penicillum* meaning a brush and the suffix *-atus* meaning provided with, alludes to the long brushlike setules of the modified seta on the second maxilliped segment of the male.

Remarks.—*Doridicola penicillatus* may be distinguished from its congeners by two easily observed characters, the pair of long slender spiniform processes on the genital segment of the female, and the brushlike modification of the seta on the second segment of the maxilliped of the male. No other species in the genus shows these features.

Doridicola dunnae sp. n.
Figs. 19a–j, 20a–j

Type material.—2 ♀♀ from one sea anemone, *Heteractis crispa* (Ehrenberg), in 3 m, north of Isle Maître, near Nouméa, New Caledonia, 22°19′30″S, 166°24′35″E, 13 July 1971. (This is the same host individual from which *Doridicola cylichnophorus, D. paterellis,* and *D. scyphulanus* were recovered.) Holotype ♀ (dissected) deposited in the National Museum of Natural History, Smithsonian Institution, Washington, D.C.: one paratype (dissected) in the collection of the author.

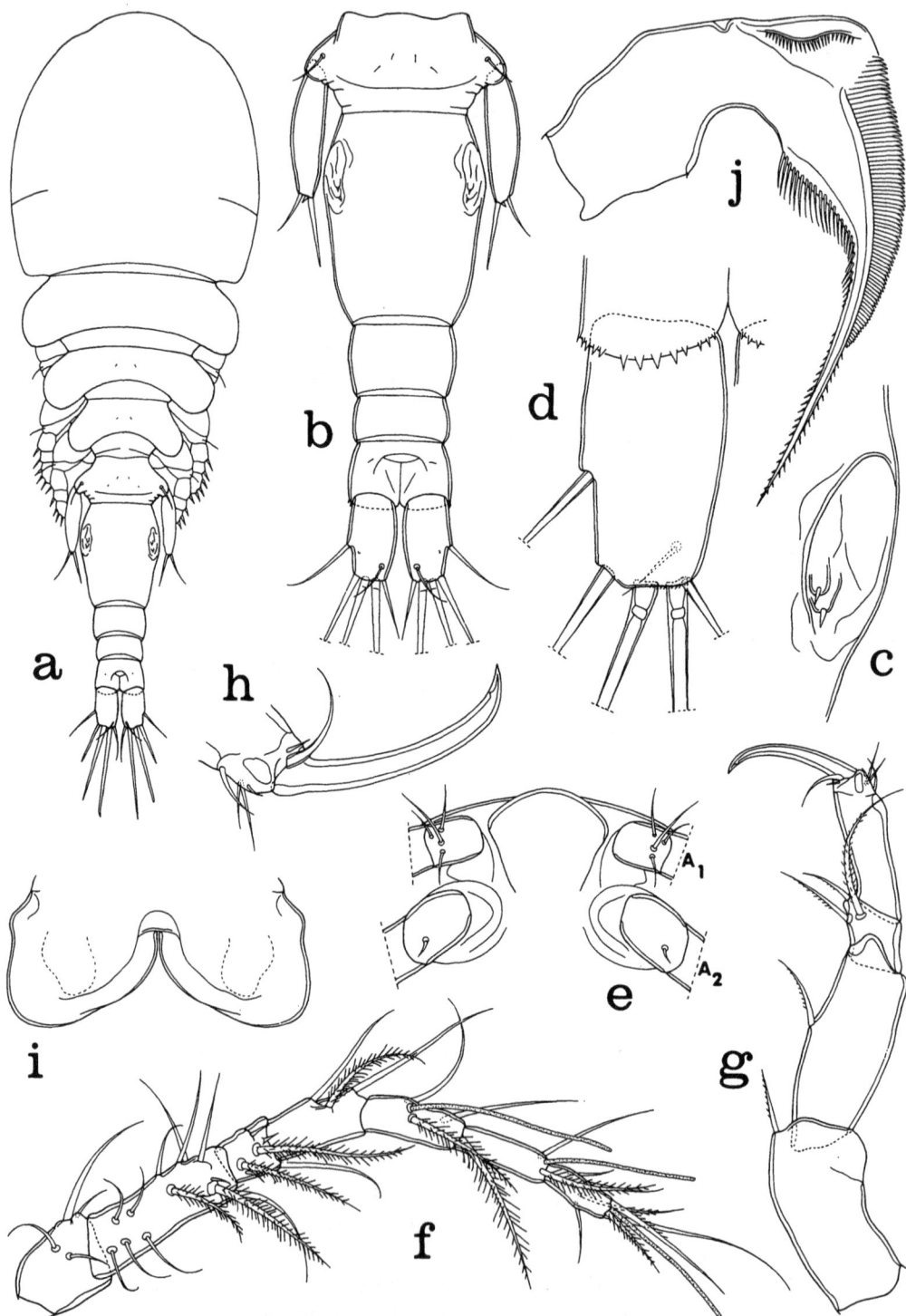

FIG. 19. *Doridicola dunnae* sp. n., female: a, dorsal (A); b, urosome, dorsal (B); c, genital area, dorsal (C); d, caudal ramus, ventral (F); e, rostrum, ventral (E); f, first antenna, ventro-outer (D); g, second antenna, posterior (D); h, tip of fourth segment of second antenna, posterior (C); i, labrum, with paragnaths indicated by broken lines, ventral (D); j, mandible, posterior (C).

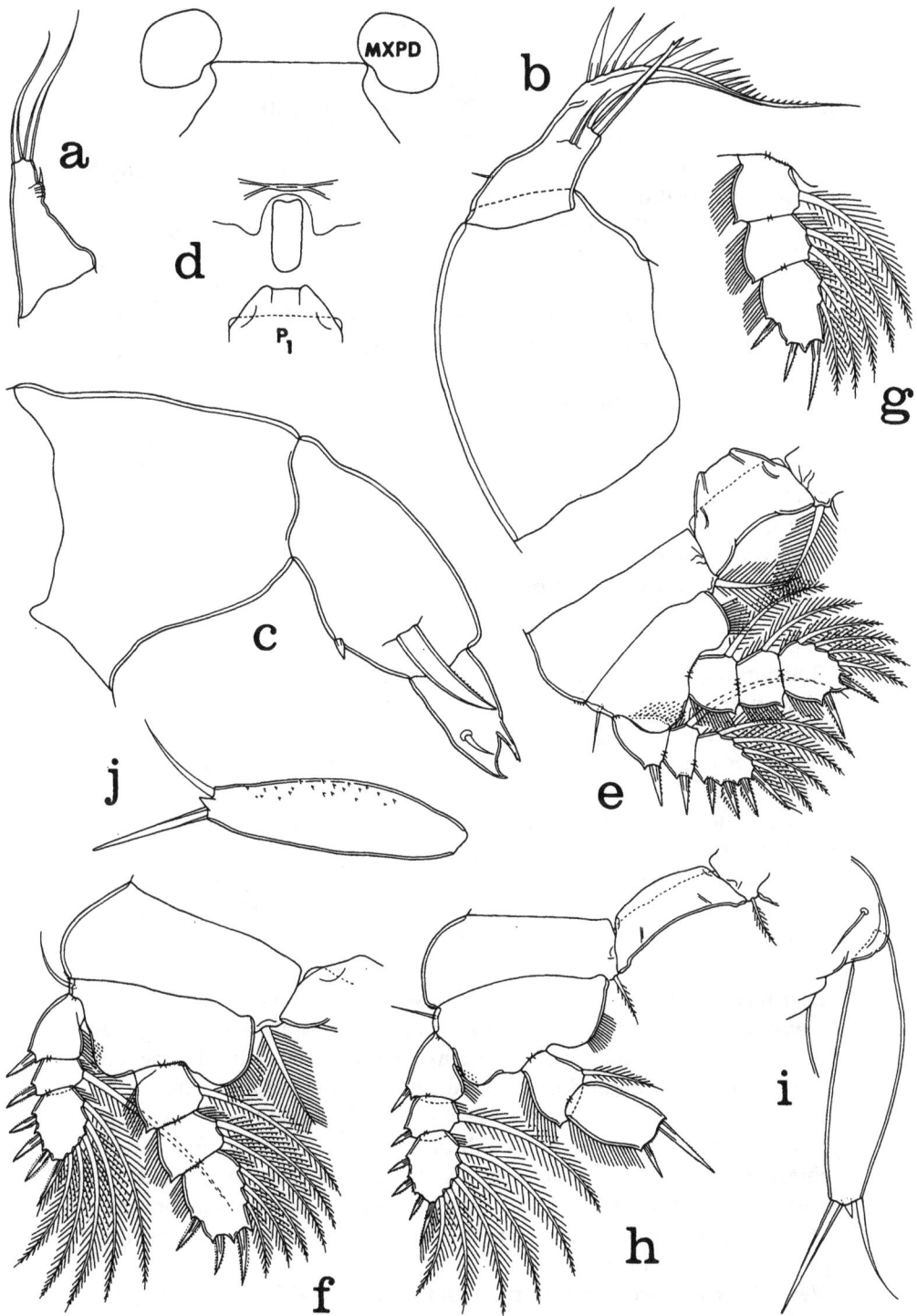

FIG. 20. *Doridicola dunnae* sp. n., female: a, first maxilla posterior (C); b, second maxilla, posterior (C); c, maxilliped, postero-inner (C); d, area between maxillipeds and first pair of legs, ventral (E); e, leg 1 and intercoxal plate, anterior (E); f, leg 2, anterior (E); g, endopod of leg 3, anterior (E); h, leg 4 and intercoxal plate, anterior (E); i, leg 5, dorsal (D); j, free segment of leg 5, flat ventral view (D).

Female.—Body (Fig. 19a) elongate. Length and greatest width of two specimens 2.09 × 0.79 mm and 2.11 × 0.77 mm, average 2.10 × 0.78 mm. Segment of leg 1 incompletely separated from head. Epimera of segments of legs 1–4 rounded. Ratio of length to width of prosome 1.66:1. Ratio of length of prosome to that of urosome 1.62:1.

Segment of leg 5 (Fig. 19b) 150 × 308 μm. Genital segment elongate, 310 × 253 μm. Genital areas situated dorsolaterally in anterior half of segment. Each area (Fig. 19c) with two minute setae approximately 5.5 μm long. Three postgenital segments from anterior to posterior 101 × 161, 78 × 148, and 88 × 155 μm. Anal segment with posteroventral margin bearing spines (Fig. 19d).

Caudal ramus (Fig. 19d) moderately elongate, 133 × 78 μm, ratio 1.71:1. Outer lateral seta 90 μm, dorsal seta 44 μm, and innermost terminal seta 96 μm. Outermost terminal seta 90 μm, and two median terminal setae 234 μm (outer) and 273 μm (inner), all three setae with blunt tips. Two median terminal setae inserted between slight dorsal flange (smooth) and ventral flange (with row of small spines). Longest terminal seta relatively short, only a little more than two times length of ramus. All six setae smooth.

Body surface with few hairs (sensilla) as in Figure 19a.

Egg sac unknown.

Rostrum (Fig. 19e) incomplete posteroventrally.

First antenna (Fig. 19f) 489 μm long. Lengths of seven segments: 49 (73 μm along anterior margin), 127, 42, 62, 68, 68, and 49 μm respectively. Formula for armature as in preceding species of *Doridicola*. Certain setae on all segments except first segment haired on ventro-outer surfaces.

Second antenna (Fig. 19g) 4-segmented and 407 μm long including claws, with formula 1, 1, 3, and 2 claws plus 5 setules. Longest seta on third segment with surficial hairs. Fourth segment 120 μm along outer side, 83 μm along inner side, and 29 μm wide. Two claws (Fig. 19h) very unequal, longer claw strong and 96 μm along its axis, shorter claw slender and 39 μm, slightly expanded proximally, setiform distally.

Labrum (Fig. 19i) with two divergent posteroventral lobes. Mandible (Fig. 19j) with general form as in other species of *Doridicola*. Paragnath (Fig. 19i) a small, apparently smooth lobe. First maxilla (Fig. 20a) with 3 setae, one of them very short. Second maxilla (Fig. 20b) segmented and armed as in other species of *Doridicola*, two setae on second segment without lateral spinules (except for one minute tooth on longer of these setae) and teeth (spines) on lash long and slender. Maxilliped (Fig. 20c) segmented and armed as in other species of *Doridicola*. Two setae on second segment very unequal, larger seta broad with minute barbs along one side. Tip of third segment slightly hooked.

Ventral area between maxillipeds and first pair of legs (Fig. 20d) not protuberant.

Legs 1–4 (Fig. 20e–h) segmented and armed as in other species of *Doridicola* except third segment of endopod of leg 1 which has formula I, I, 4 instead of usual I, 5 as in other species. Leg 4 (Fig. 20h) with exopod 169 μm long and inner coxal seta 52 μm and finely plumose. Endopod with first segment

60 × 62 μm, its inner plumose seta 88 μm; second segment 99 × 58 μm (length 104 μm including small distal spiniform process), its two terminal minutely barbed spines 35 μm (outer) and 75 μm (inner). Outer margins of both endopod segments with hairs.

Leg 5 (Fig. 20i) with long free segment 180 × 52 μm, reaching approximately to middle of genital segment. Inner terminal seta 78 μm and outer terminal seta 87 μm (with blunt tip); small spiniform process between these two setae. Ventral outer surface of free segment with minute spines (Fig. 20j).

Leg 6 represented by two setae on genital area (Fig. 19c).

Color of living specimens in transmitted light opaque gray, eye red.

Male.—Unknown.

Etymology.—This species is named for Dr. Daphne F. Dunn of the California Academy of Sciences, who contributed in a very important way to this study by providing the identifications of the New Caledonian sea anemones.

Remarks.—*Doridicola dunnae*, of which only the female is known, exceeds in length all species except *D. ptilosarci*, 1.94 mm (1.82–2.08 mm), and *D. magnificus*, 3.06 mm (2.98–3.17 mm). The new species differs from all its congeners in having certain setae on the first antennae haired on their ventro-outer surfaces. Other distinctive features include the very unequal claws on the second antennae, the very dissimilar setae on the second segment of the maxilliped, and the unusually large spines on the posteroventral margin of the anal segment.

Doridicola titillans sp. n.
Figs. 21a–h, 22a–i, 23a–h

Type material.—2 ♀♀, 1 ♂ from two specimens of the actiniarian *Condylactis gigantea* (Weinland), in 1 m, western shore of Cabo Rojo, southwestern Puerto Rico, 24 August 1959. Holotype ♀, allotype, and one paratypic female deposited in the National Museum of Natural History, Smithsonian Institution, Washington, D.C.

Other specimens.—2 ♀♀ from one *Condylactis gigantea*, in 1 m, La Pelota, near La Cueva, southwestern Puerto Rico, 14 August 1959.

Female.—Body (Fig. 21a) elongate, slender, and strongly flexed in specimens preserved in 70 per cent ethyl alcohol (Fig. 21b). Length 2.18 mm (2.10–2.27 mm), greatest width 0.65 mm (0.62–0.67 mm), and dorsoventral thickness of prosome 0.55 mm, based on four specimens. Epimera of segments of legs 1–4 rounded posteriorly. Ratio of length to width of prosome 1.93:1. Ratio of length of prosome to that of urosome 1.16:1.

Segment of leg 5 (Fig. 21c) 143 × 325 μm. Genital segment 320 × 275 μm, broadest anteriorly and tapered posteriorly. Genital areas situated dorsolaterally in anterior third. Each area (Fig. 21d) with two naked setae 13 μm and 16.5 μm. Three postgenital segments from anterior to posterior 165 × 169, 154 × 144, and 126 × 136 μm. Posteroventral margin of anal segment smooth.

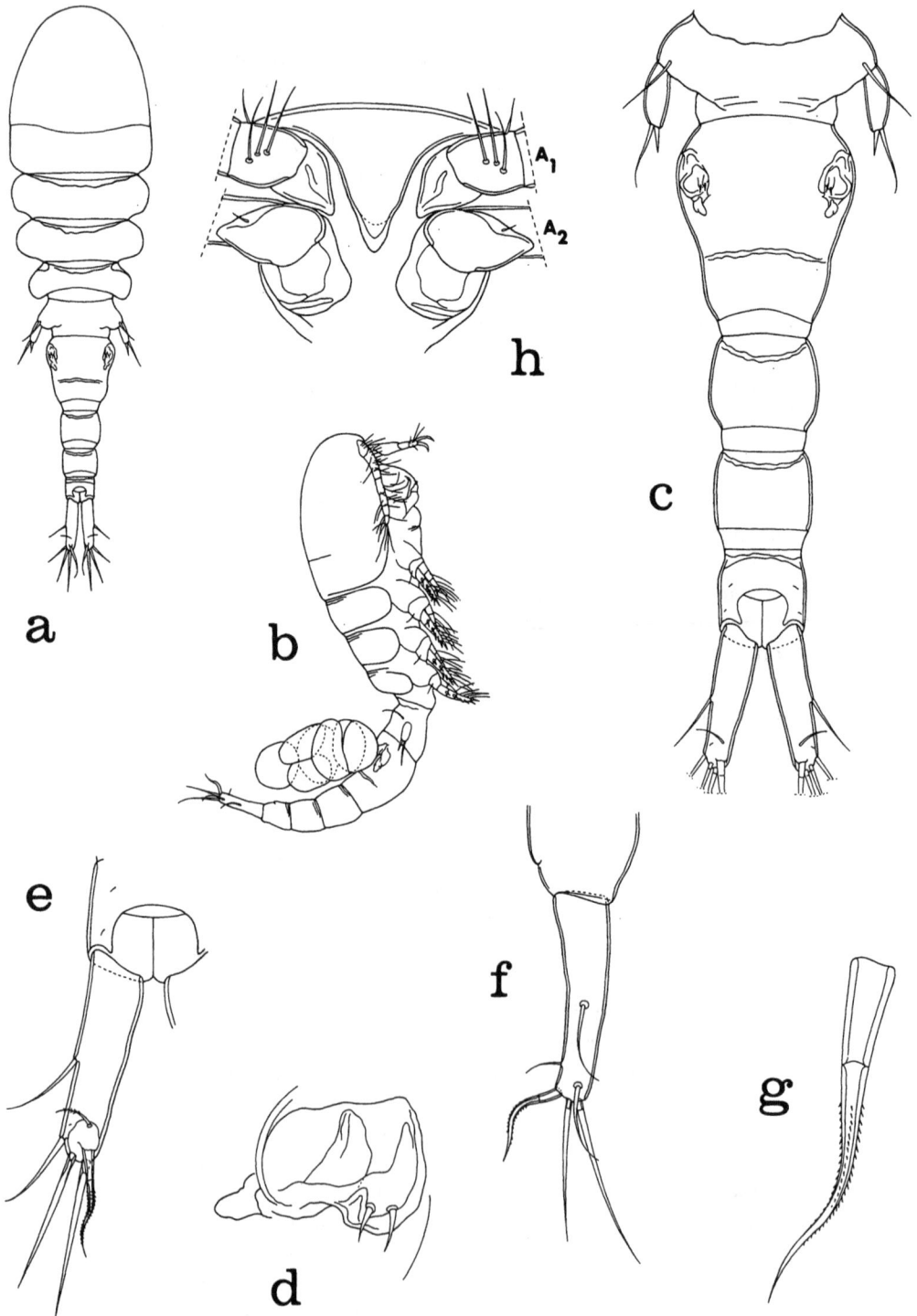

FIG. 21. *Doridicola titillans* sp. n., female: a, dorsal (I); b, lateral (I); c, urosome, dorsal (B); d, genital area, lateral (C); e, caudal ramus, dorsal (E); f, caudal ramus, lateral (E); g, innermost terminal seta on caudal ramus, dorsal (G); h, rostrum, ventral (E).

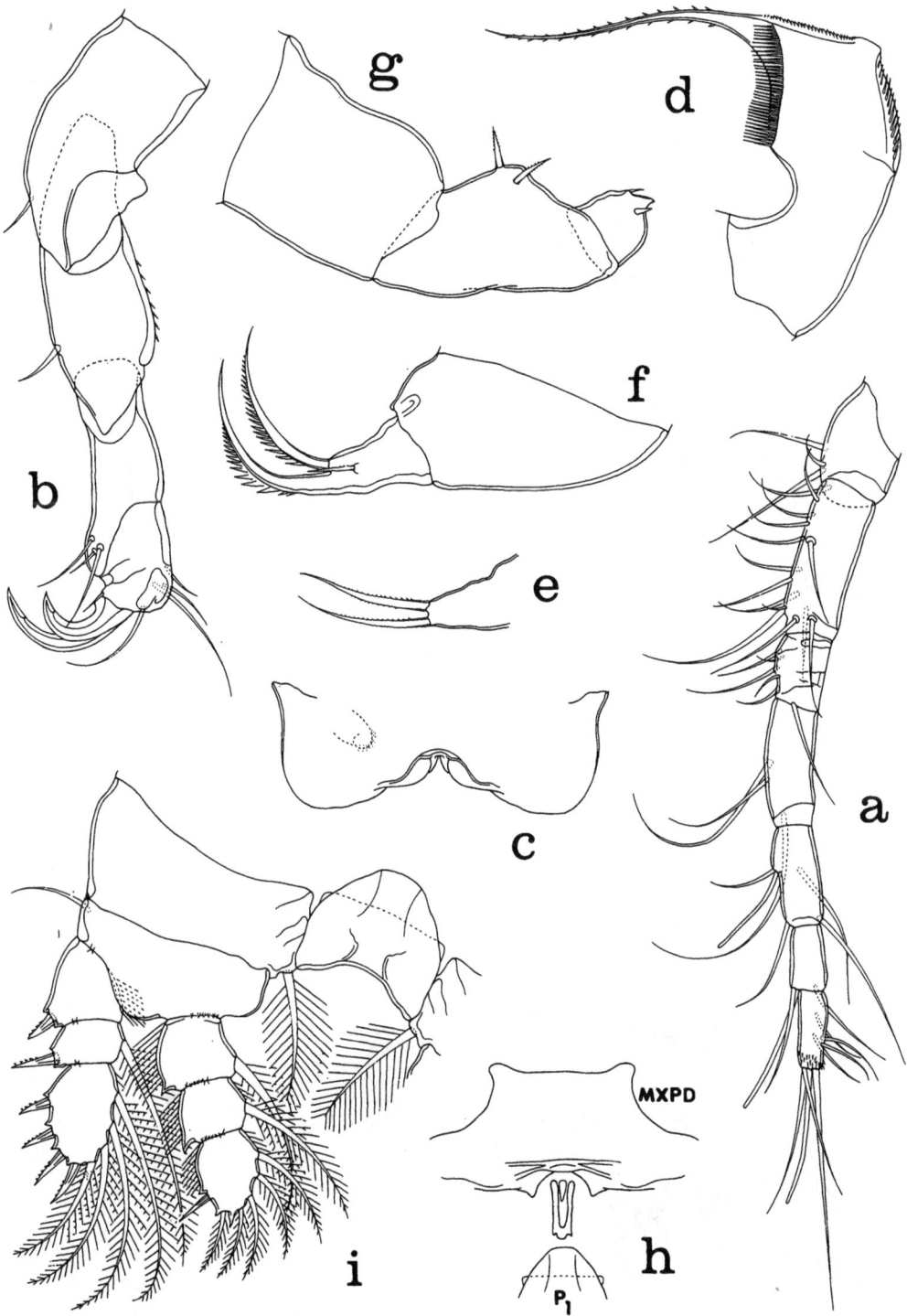

FIG. 22. *Doridicola titillans* sp. n., female: a, first antenna, ventral (D); b, second antenna, posterior (F); c, labrum, with paragnaths indicated by broken lines, ventral (F); d, mandible, posterior (G); e, first maxilla, anterior (G); f, second maxilla, posterior (C); g, maxilliped, postero-inner (C); h, area between maxillipeds and first pair of legs, ventral (E); i, leg 1 and intercoxal plate, anterior (D).

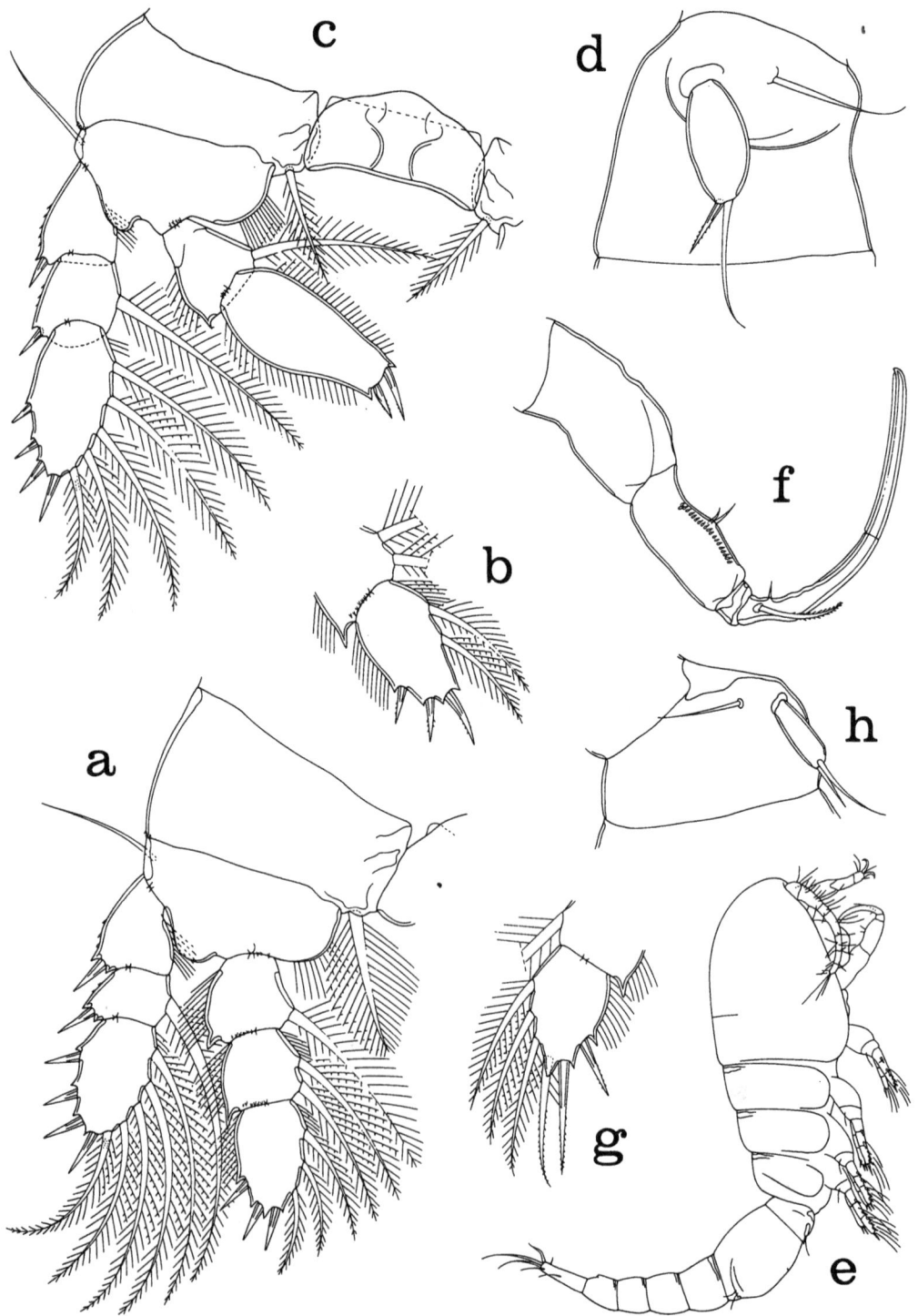

FIG. 23. *Doridicola titillans* sp. n. Female: a, leg 2, anterior (D); b, third segment of endopod of leg 3, anterior (D); c, leg 4 and intercoxal plate, anterior (D); d, segment of leg 5 and 5, lateral (D). Male: e, lateral (A); f, maxilliped, postero-inner (D); g, third segment of endopod of leg 1, anterior (F); h, segment of leg 5 and leg 5, lateral (D).

Caudal ramus (Fig. 21e, f) elongate, 211 × 55 μm, ratio 3.84:1. Outer lateral seta 90 μm and naked. Dorsal seta 40 μm and delicately haired. Outermost terminal seta 87 μm and naked. Innermost terminal seta 99 μm, stout, well sclerotized with proximal "joint," slightly sigmoid or whiplike, standing erect (Fig. 21f), and bearing three rows of spinules, two lateral rows and one median row (Fig. 21g). Two smooth median terminal setae 148 μm (outer) and 174 μm (inner).

Body surface mostly devoid of hairs (sensilla).

Egg sac (Fig. 21b) elongate oval, 560 × 264 μm, containing seven irregular eggs with dimensions ranging from 132 to 253 μm.

Rostrum (Fig. 21h) narrow and pointed posteroventrally.

First antenna (Fig. 22a) 465 μm long. Length of seven segments: 34 (70 μm along anterior side), 96, 57, 73, 70, 47, and 52 μm respectively. Formula for armature as in preceding species of Doridicola. All setae naked. Distal part of last segment ornamented with ventral group of spinules extending around partly on dorsal side of segment.

Second antenna (Fig. 22b) 277 μm long without claws and 4-segmented. Formula for armature as in preceding species of Doridicola. Two claws 47 μm and 65 μm. All setae smooth. Outer side of second segment ornamented with row of spinules.

Labrum (Fig. 22c) with two broad short lobes. Mandible (Fig. 22d) with convex margin of base somewhat angular. Paragnath (Fig. 22c) a small hairy lobe. First maxilla (Fig. 22c) with two finely barbed setae. Second maxilla (Fig. 22f) with inner seta and lash of approximately equal length. Maxilliped (Fig. 22g) segmented and armed as in preceding species of Doridicola; two barbed setae on second segment nearly equal.

Ventral area between maxillipeds and first pair of legs (Fig. 22h) slightly protuberant (Fig. 21b).

Legs 1-4 (Figs. 21i, 23a-c) segmented and armed as in species of Doridicola described above. Exopod spines on legs 2-4 apparently smooth. Leg 4 (Fig. 23c) with exopod 216 μm long. First segment of endopod 62 × 60 μm, its inner plumose seta 138 μm; second segment 130 × 60 μm in greatest dimensions, its terminal minutely barbed spines 29 μm (outer) and 43 μm (inner), and inner margin with row of hairs. Inner coxal seta 83 μm.

Leg 5 (Fig. 23d) with unornamented free segment 81 × 45 μm, ratio 1.8:1. Two terminal setae 37 μm (finely barbed) and 91 μm (smooth). Dorsal seta 99 μm and naked.

Leg 6 represented by two setae on genital area (Fig. 21d).

Color unknown.

Male.—Body (Fig. 23e) resembling that of female. Length of allotype 1.78 mm and greatest width 0.53 mm. Genital segment elongate. Four postgenital segments. Caudal ramus as in female.

Rostrum as in female. First antenna like that of female, without added aesthetes. Second antenna, labrum, mandible, paragnath, first maxilla, and second maxilla like those of female (insofar as could be determined without dissection). Maxilliped (Fig. 23f) segmented and armed as in previously described species of *Doridicola*. First segment elongate and unarmed. Second segment with two small inner naked setae and row of spines. Small third segment unarmed. Claw 214 μm along its axis, without terminal lamella, subdivided midway along its length, and bearing proximally two very unequal setae, shorter seta smooth, longer seta distally barbed. Ventral area between maxillipeds and first pair of legs as in female.

Legs 1–4 segmented and armed as in female, except for endopod of leg 1 where third segment (Fig. 23g) has formula I, I, 4, with outermost seta plumose proximally and barbed distally.

Leg 5 (Fig. 23h) with free segment 49 × 20 μm, its terminal setae 36 μm and 65 μm. Dorsal seta 55 μm.

Leg 6 a posteroventral flap on genital segment bearing two small setae. Spermatophore not seen.

Color unknown.

Etymology.—The specific name *titillans*, from Latin *titillo* meaning to tickle, alludes in a fanciful way to the erect seta on the caudal ramus.

Remarks.—*Doridicola titillans* differs from most species in the genus by its large size. In only three other species, *Doridicola magnificus* (Humes, 1964), *Doridicola ptilosarci* Humes and Stock, 1973, and *Doridicola dunnae* described above, does the length of the female exceed 2 mm. Females of the three species may be easily distinguished from *D. titillans* as follows: the caudal ramus of *D. magnificus* is much longer (ratio 7.8:1), and the caudal ramus of *D. ptilosarci* and *D. dunnae* is shorter (1.5:1 and 1.7:1 respectively).

The slender body of *D. titillans* separates the species from most congeners. The erect sigmoid seta on the caudal ramus is unique in the genus and serves as a good recognition character. With only two setae on the first maxilla *D. titillans* differs from all other members of the genus where, as far as known, the first maxilla has either three or four setae.

Keys to the species of *Doridicola* associated with Actiniaria (*Doridicola antheae* Ridley, 1879, not included on account of its uncertain identity.)

Females

1. Length of body 2.98–3.17 mm; caudal ramus elongate, 7.8:1
   .... *D. magnificus*
   Length of body less than 2.5 mm; caudal ramus less than 4:1 2
2. With prominent lobe adjacent to insertion of free segment of
   leg 5 3
   Without such lobes (small mammilliform lobe with acuminate tip in
   D. penicillatus) 4
3. Lobe adjacent to leg 5 bearing several small spines; genital area without
   patch of small spines; genital segment much longer than wide
   .... *D. caelatus*
   Lobe adjacent to leg 5 smooth; genital area with patch of small spines;
   genital segment wider than long .... *D. hispidulus*
4. With pair of long slender spiniform processes on genital segment
   D. penicillatus

Without pair of long slender spiniform processes on genital segment 5

5. Claws on second antenna very unequal, one slender and almost setiform; several setae on first antenna with hairs on ventro-outer surfaces .... *D. dunnae*
   Claws on second antenna nearly equal or if distinctly unequal both stout and unguiform; setae on first antenna generally smooth 6

6. First maxilla with four setae 7
   First maxilla with two or three setae 8

7. Caudal ramus 2.17:1; genital segment distinctly expanded in anterior half; claws on second antenna nearly equal .... *D. actiniae*
   Caudal ramus 1.8:1; genital segment only slightly wider in anterior half than posteriorly; claws on second antenna unequal, one claw about twice as long as other .... *D. gemmatus*

8. First maxilla with two setae; innermost terminal seta on caudal ramus sigmoid and erect .... *D. titillans*
   First maxilla with three setae; innermost terminal seta on caudal ramus straight and not erect 9

9. Caudal ramus 2.36:1; second antenna without small spinules on first three segments, fourth segment relatively short and stout, 3.3:1 .... *D. cylichnophorus*
   Caudal ramus about 1.7:1; second antenna with small spinules on inner margin of first three segments, fourth segment relatively long and slender, approximately 7:1 10

10. Genital segment with sides nearly parallel but having very slight hourglass form; leg 5 about as long as genital segment, ratio 6:1; egg sac oval .... *D. scyphulanus*
    Genital segment broadest in anterior third and tapering posteriorly; leg 5 shorter than genital segment, ratio 3.83:1; egg sac elongate *D. paterellis*

Males

(Male of *D. dunnae* is unknown.)

1. Length of body 2.62–2.98 mm .... *D. magnificus*
   Length of body not exceeding 1.8 mm 2

2. First maxilla with four setae 3
   First maxilla with two or three setae 4

3. Third segment of endopod of leg 1 with two terminal processes very unequal, longer process minutely barbed along inner edge and having minute fingerlike tip .... *D. gemmatus*
   Third segment of endopod of leg 1 with two terminal processes subequal with granular surfaces .... *D. actiniae*

4. First maxilla with two setae; innermost terminal seta on caudal ramus sigmoid and erect .... *D. titillans*
   First maxilla with three setae; innermost terminal seta on caudal ramus

straight and not erect 5

5. With suckers on second antenna 6

Without suckers on second antenna 8

6. Second antenna with suckers only on second segment, fourth segment relatively short and stout . . . . *D. cylichnophorus*

Second antenna with suckers on both first and second segments, fourth segment relatively long and slender 7

7. Second antenna with about 26 suckers on second segment and with small spinules along inner margin of fourth segment; claw of maxilliped not swollen proximally . . . . *D. paterellis*

Second antenna with about 10 suckers on second segment and inner margin of fourth segment smooth; claw of maxilliped swollen proximally . . . . *D. scyphulanus*

8. Second antenna with saucerlike rings on first two segments . . . . *D. caelatus*

Second antenna with spinules on inner surface of all four segments 9

9. Caudal ramus 2.97:1; one of two setae on second segment of maxilliped with long brushlike modification . . . . *D. penicillatus*

Caudal ramus 1.77:1; two setae on second segment of maxilliped unmodified . . . . *D. hispidulus*

Genus *Indomolgus* Humes and Ho, 1966

Indomolgus panikkari (Gnanamuthu, 1955)
= *Lichomolgus panikkari* Gnanamuthu, 1955

Host: *Phytocoeteopsis ramunni* Panikkar.
Site: Gastral cavity.
Locality: Madras, India.
Notes: Length of ♀ 2.9 mm, ♂ 2.7 mm.
Citations: Bouligand (1966), Humes and Ho (1966).

Lambanetes gen. n.

Diagnosis.—Lichomolgidae. Body elongate. Urosome of female 5-segmented, that of male 6-segmented. Caudal ramus with six setae. Rostrum not developed. First antenna 7-segmented, fourth segment with only two setae. Second antenna 3-segmented with two terminal claws.

Labrum with median posteroventral process. Mandible beyond constriction having on convex side a prominent spine followed by several toothlike spines and on concave side a row of spinules. Lash relatively short, spinelike, with strong barbs. Paragnath a small lobe with hairs. First maxilla with two setae. Second maxilla 2-segmented. Maxilliped of male 4-segmented if claw is considered as a segment.

Legs 1–3 with 3-segmented rami. Leg 4 with 3-segmented exopod but 2-segmented endopod. Inner coxal seta absent in all four legs. Terminal segments of exopods of legs 2–4 and of endopods of legs 2 and 3 with reduced armature. Leg 4 endopod with 0-1; II.

Leg 5 placed lateroventrally, free segment bearing two terminal setae.

Other features as in species described below.

Associated with actiniarians.

Gender masculine.

Type-species.—Lambanetes stichodactylae sp. n.

Etymology.—The generic name is formed from λαμβανω, meaning to grasp, and the suffix -της, signifying action or agency, alluding to the powerful maxillipeds.

The genus *Lambanetes* differs from other lichomolgid genera in the combination of several features: (1) the presence of only two instead of three setae on the fourth segment of the first antenna, (2) the 3-segmented second antenna with two terminal claws, (3) the median process on the labrum, (4) the absence of an inner coxal seta on all four legs, (5) the reduced number of spines and setae on the terminal exopod segments of legs 2–4 and the terminal endopod segments of legs 2 and 3.

Lambanetes stichodactylae sp. n.
Figs. 24a–e, 25a–k, 26a–g

Type material.—2 ♂♂ from one individual of the actiniarian *Stichodactyla haddoni* (Saville-Kent), in tide pool, Ricaudy Reef, near Nouméa, New Caledonia, 22°19′00″S, 166°26′44″E, 21 July 1971. Holotype deposited in the National Museum of Natural History, Smithsonian Institution, Washington, D.C.; paratype (dissected) in the collection of the author.

Other material.—1 ♂ from one individual of the actiniarinan *Stichodactyla gigantea* (Forskål), intertidal, Ricaudy Reef, near Nouméa, New Caledonia, 21 July 1971.

Male.—Body (Fig. 24a, b) moderately elongate, with cephalosome only a little wider than metasome. Length of body 1.12 mm (0.96–1.22 mm) and greatest width 0.60 mm (0.58–0.63 mm), based on three specimens. Ratio of length to width of prosome 1.18:1. Ratio of length of prosome to that of urosome 1.37:1. Epimeral areas of metasomal segments as in Figure 24b.

Segment of leg 5 (Fig. 24c) short, 34 × 280 µm. Genital segment broad, 156 × 335 µm. Four postgenital segments from anterior to posterior 60 × 161, 60 × 148, 52 × 130, and 72 × 109 µm. Posteroventral margin of anal segment smooth.

Caudal ramus (Fig. 24d) 49 × 29 µm in greatest dimensions. Outer lateral seta 26 µm, dorsal seta 57 µm, outermost terminal seta 29 µm, innermost terminal seta 70 µm, and two median terminal setae 47 µm (outer) and 208 µm (inner). All setae smooth.

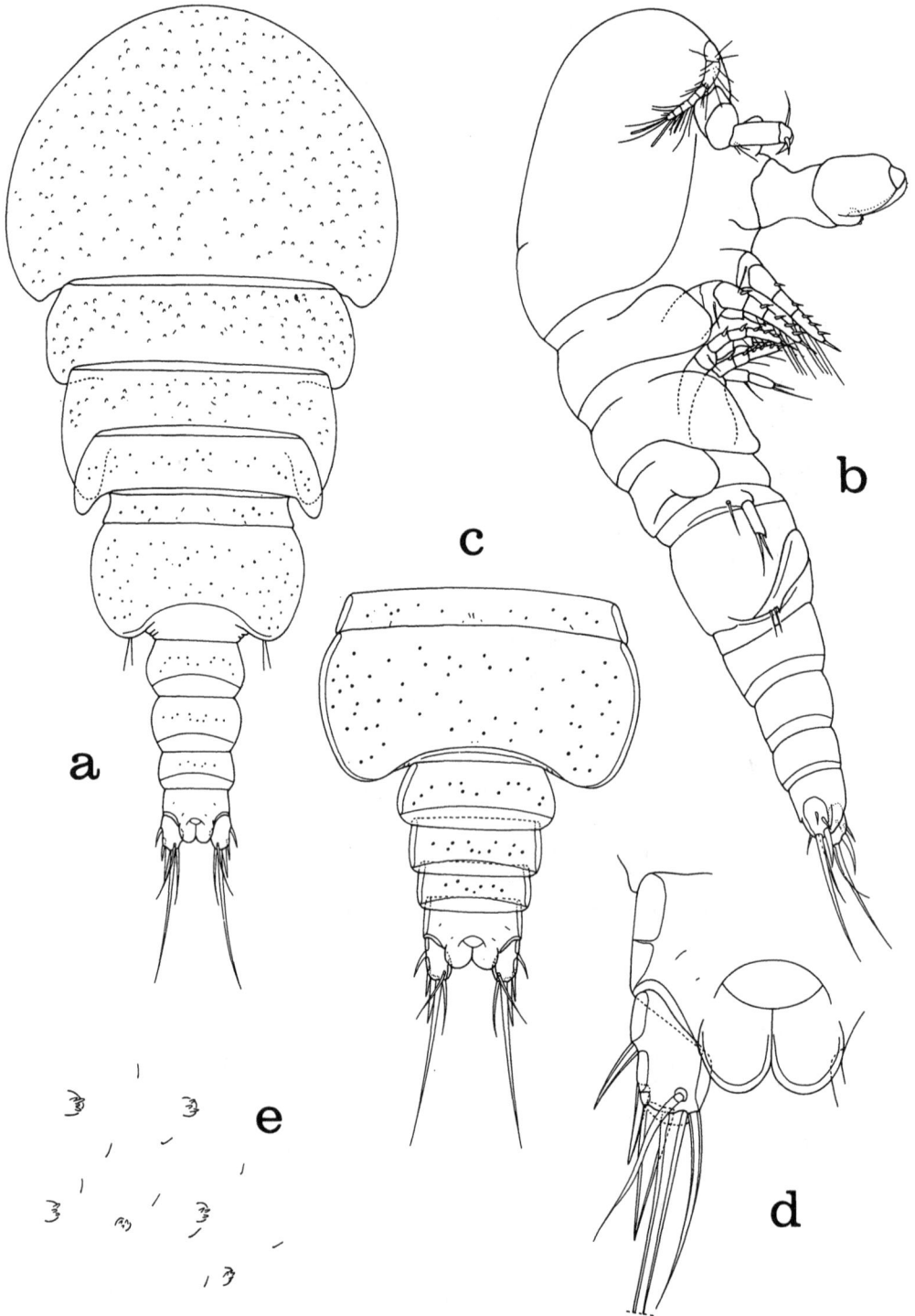

FIG. 24. *Lambanetes stichodactylae* gen. n., sp. n., male: a, dorsal (B); b, lateral, fine ornamentation of body omitted (B); c, urosome, dorsal (F); d, caudal ramus, dorsal (C); e, portion of dorsal surface of head, dorso–anterior (H).

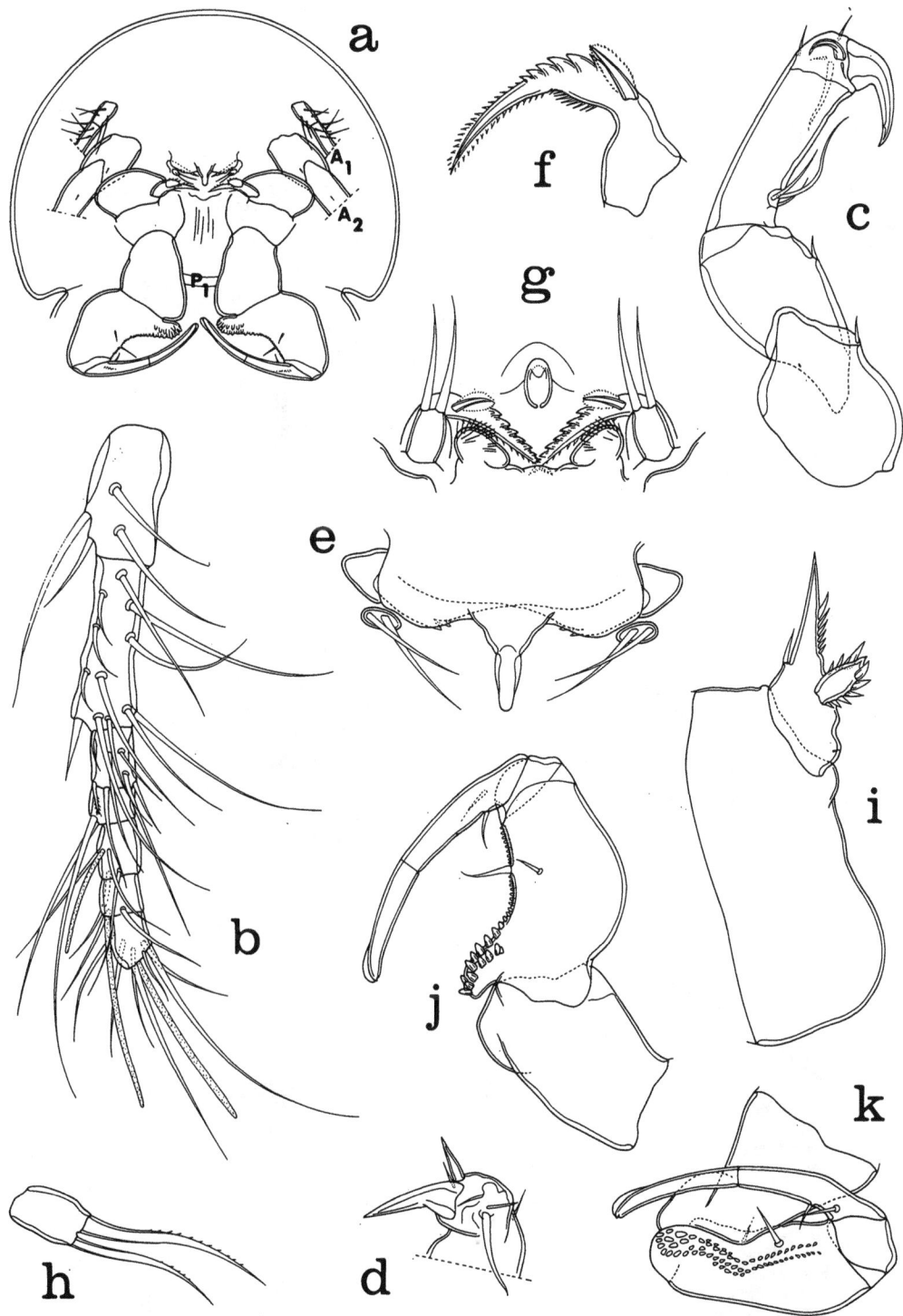

FIG. 25. *Lambanetes stichodactylae* gen. n., sp. n., male: a, head, ventral (B); b, first antenna, ventral (C); c, second antenna, postero-inner (F); d, tip of fourth segment of second antenna, end view (F); e, labrum, with mandibles and first maxillae, ventral (C); f, mandible, posterior (G); g, mouth area, showing mandibles, paragnaths, and first maxillae, posterior, with labrum turned forward (C); h, first maxilla, posterior (G); i, second maxilla, anterior (C); j, maxilliped, postero-inner (D); k, maxilliped, inner (D).

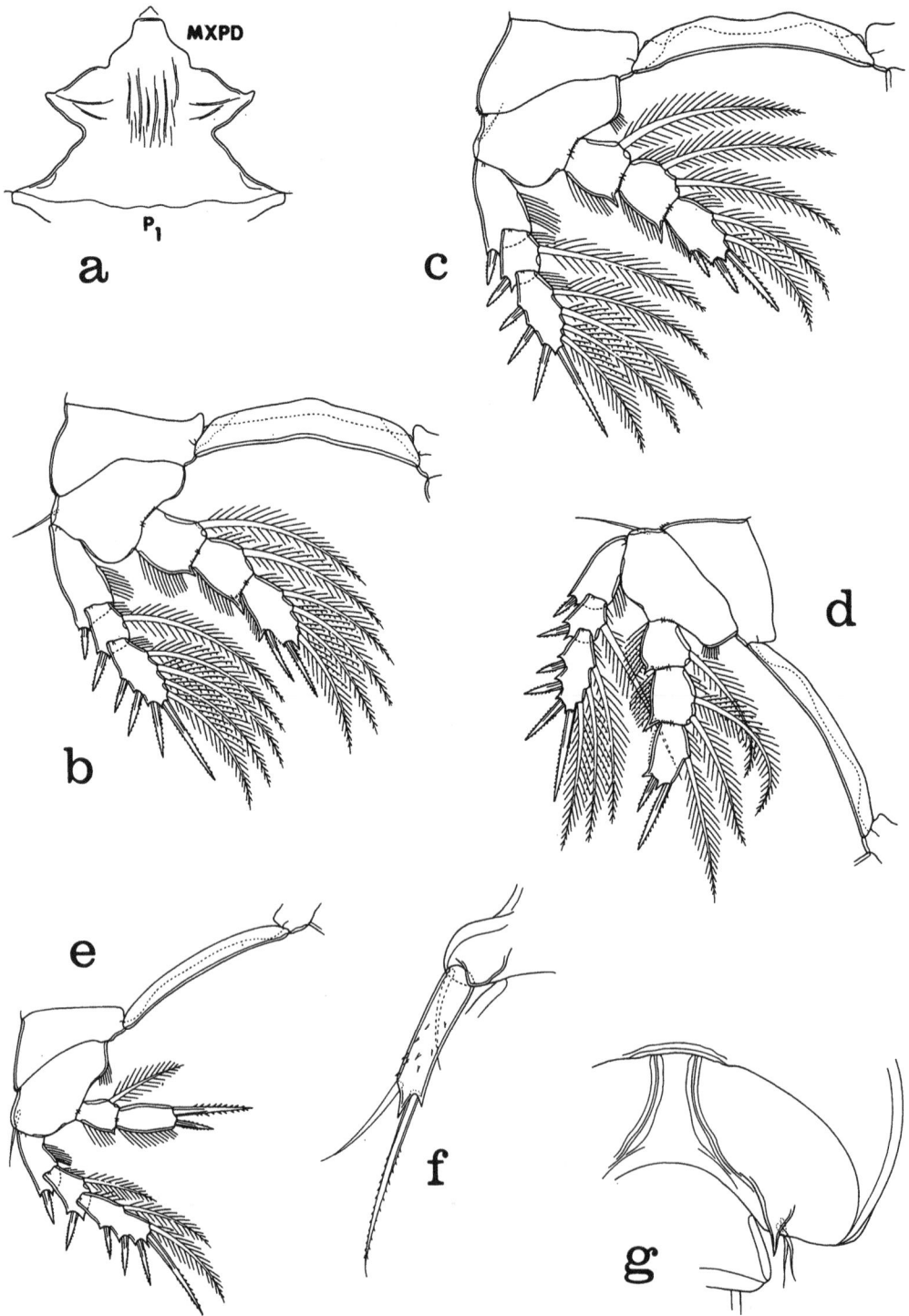

FIG. 26. *Lambanetes stichodactylae* gen. n., sp. n., male: a, area between maxillipeds and first pair of legs, ventral (D); b, leg 1 and intercoxal plate, anterior (D); c, leg 2 and intercoxal plate, anterior (D); d, leg 3 and intercoxal plate, anterior (D); e, leg 4 and intercoxal plate, anterior (D); f, leg 5, ventral (F); g, leg 6, ventral (D).

Body surface profusely covered dorsally with sensory knobs and hairs (sensilla) as in Figure 24a. Each sensory knob with two or three extremely minute short processes (Fig. 24e).

Rostrum (Fig. 25a) not developed. First antenna (Fig. 25b) 189 μm long. Lengths of seven segments: 48 (50 μm along anterior margin), 60, 19, 11, 18, 12, and 19 μm respectively. Formula for armature: 4, 13, 6, 2, 4, + 1 aesthete, 2 + 1 aesthete, and 7 + 1 aesthete. All setae naked, except for minute barbs on short seta on third segment.

Second antenna (Fig. 25c) 3-segmented, 195 μm long. Formula: 1, 1, 3, + 2 claws and 5 setules. Claws recurved (but may appear nearly straight as in Figure 25d), larger claw 47 μm, smaller claw about 20 μm. All setae smooth.

Labrum (Fig. 25e) broad with median posteroventral process. Mandible (Fig. 25f, g) small, 60 μm long, with convex side beyond indentation bearing a spine with serrate lamellae and several toothlike spines, and with concave side having row of spinules. Lash moderately short, spiniform, and barbed. Paragnath (Fig. 25g) a small lobe with hairs. First maxilla (Fig. 25g, h) with two setae. Second maxilla (Fig. 25i) 2-segmented. First segment short and broad, gibbous on outer side. Second segment with outer surficial seta swollen and bearing strong lateral spines, inner seta slender and smooth; segment terminating in a short spiniform lash with spines along one side. Maxilliped (Fig. 25j, k) large, 4-segmented, assuming that proximal part of claw represents fourth segment. First segment unarmed. Second segment with two smooth setae and spines along inner surface, these spines becoming larger on inner proximal protrusion of segment. Small third segment unarmed. Claw 200 μm along its axis, with small terminal lamella. Claw subdivided midway along its length and bearing two small proximal smooth setae.

Ventral area between maxillipeds and first pair of legs (Fig. 26a) not protuberant and lacking median sclerite usually seen in lichomolgids in front of intercoxal plate of leg 1.

Legs 1–4 (Fig. 26b–e) with 3-segmented rami except for 2-segmented endopod of leg 4. Spine and setal formula as follows (Roman numerals representing spines, Arabic numerals indicating setae):

P_1	coxa	0-0	basis	1-0	exp	I-0;	I-1;	III, I, 4
					enp	0-1;	0-1;	I, I, 4
P_2	coxa	0-0	basis	1-0	exp	I-0;	I-1;	III, I, 4
					enp	0-1;	0-2;	I, II, 2
P_3	coxa	0-0	basis	1-0	exp	I-0;	I-1;	III, I, 3
					enp	0-1;	0-2;	II, 1
P_4	coxa	0-0	basis	1-0	exp	I-0;	I-1;	III, I, 2
					enp	0-1;	II	

Legs 1–4 without an inner coxal seta. Terminal segments of rami with reduced armature. Leg 4 exopod 125 μm. First segment of endopod 29 ×

23 μm with inner plumose seta 62 μm. Second segment of 33 × 21 μm with two terminal barbed spines 25 μm (outer) and 54 μm (inner).

Leg 5 (Fig. 26f) placed ventrolaterally, and thus concealed in dorsal view of urosome. Free segment 65 μm (77 μm with spiniform processes) × 13 μm wide. Two terminal setae 62 μm (smooth) and 93 μm (finely barbed) with dorsal seta about 50 μm. Free segment ornamented ventrally with few small spinules.

Leg 6 (Fig. 26g) a posteroventral flap on genital segment bearing two naked slender setae about 45 μm.

Spermatophore not seen.

Color in living specimens in transmitted light opaque gray, eye red.

Female unknown.

Etymology.—The specific name *stichodactylae* is formed from that of the host genus.

Lambanetes gemmulatus sp. n.
Figs. 27a–m, 28a–h, 29a–f

Type material.—12 ♀♀, 46 ♂♂, and 7 copepodids from one actiniarian, *Cryptodendron adhaesivum* Klunzinger, in 1 m, west of Isle Mando, near Nouméa, New Caledonia, 22°18′59″S, 166°09′30″E, 3 July 1971. Holotype ♀, allotype, and 51 paratypes (8 ♀♀, 43 ♂♂) deposited in the National Museum of Natural History, Smithsonian Institution, Washington, D.C.: the remaining paratypes (dissected) in the collection of the author.

Female.—Body (Fig. 27a) with moderately stout prosome; urosome in preserved specimens strongly flexed. Length of body 1.16 mm (1.06–1.20 mm) and dorsoventral thickness of prosome 0.32 mm (0.28–0.36 mm), based on 10 specimens. Width of prosome approximately 0.35 mm. Ratio of length to width of prosome approximately 3.31:1. Ratio of length of prosome to that of urosome about 1.32:1. Segment of leg 1 fused with cephalosome. Epimera of segments of legs 1 and 4 rounded, those of legs 2 and 3 with budlike knobs (Fig. 27b).

Segment of leg 5 (Fig. 27c) dorsally short and obscured in dorsal view by tergum of segment of leg 4, dimensions 32 × approximately 128 μm dorsally, 159 × 130 μm ventrally. Genital segment roughly quadrate, 130 × 143 μm. Genital areas located dorsally in anterior half of segment. No setae visible on these areas. Three postgenital segments from anterior to posterior 79 × 117, 65 × 105, and 75 × 94 μm (greatest length). Posteroventral margin of anal segment smooth.

Caudal ramus (Fig. 27d) 47 × 24 μm, ratio 2:1. Outer lateral seta 15 μm. Dorsal seta 23 μm. Outermost terminal seta 17 μm and innermost terminal seta 35 μm. Two median terminal setae differing markedly in appearance, outer 27 μm and stout, inner 52 μm and slender. All setae naked. Dorsal surface of ramus with two minute hairs.

Body surface smooth, without sensilla (Fig. 27a).

Egg sac not seen.

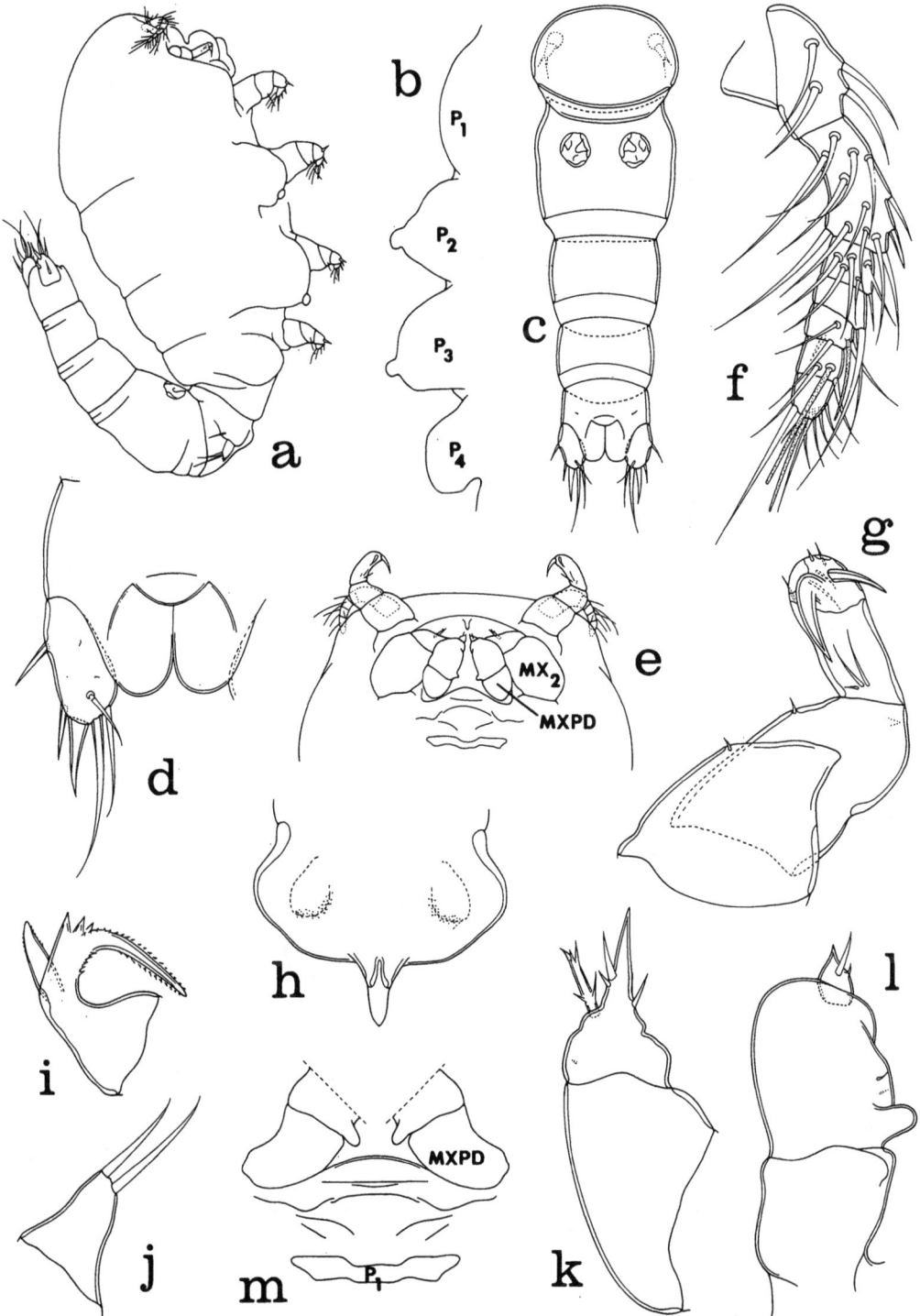

FIG. 27. *Lambanetes gemmulatus* gen. n., sp. n., female: a, lateral (B); b, contour of epimera of segments of legs 1–4, ventral (E); c, urosome, dorsal (E); d, caudal ramus, ventral (C); e, head, ventral (E); f, first antenna, anterodorsal (G); g, second antenna, postero-inner (G); h, labrum, with paragnaths indicated by broken lines, ventral (G); i, mandible, anterior (H); j, first maxilla, ventral (H); k, second maxilla, posterior (G); l, maxilliped, posterior (G); m, area between maxillipeds and first pair of legs, ventral (F).

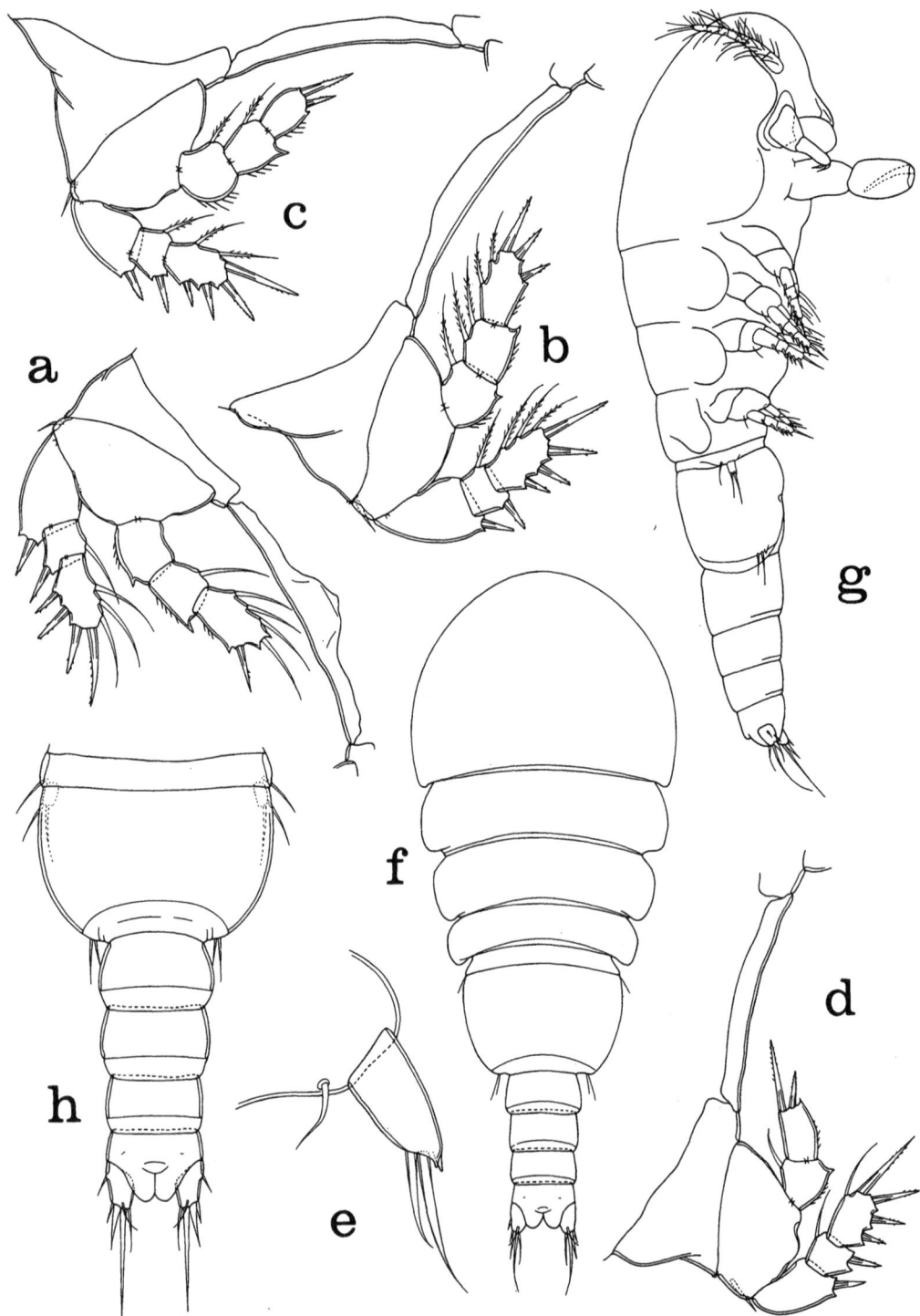

FIG. 28. *Lambanetes gemmulatus* gen. n., sp. n. Female: a, leg 1 and intercoxal plate, anterior (C); b, leg 2 and intercoxal plate, anterior (C); c, leg 3 and intercoxal plate, anterior (C); d, leg 4 and intercoxal plate, anterior (C); e, leg 5, lateral (G). Male: f, dorsal (E); g, lateral (E); h, urosome, dorsal (D).

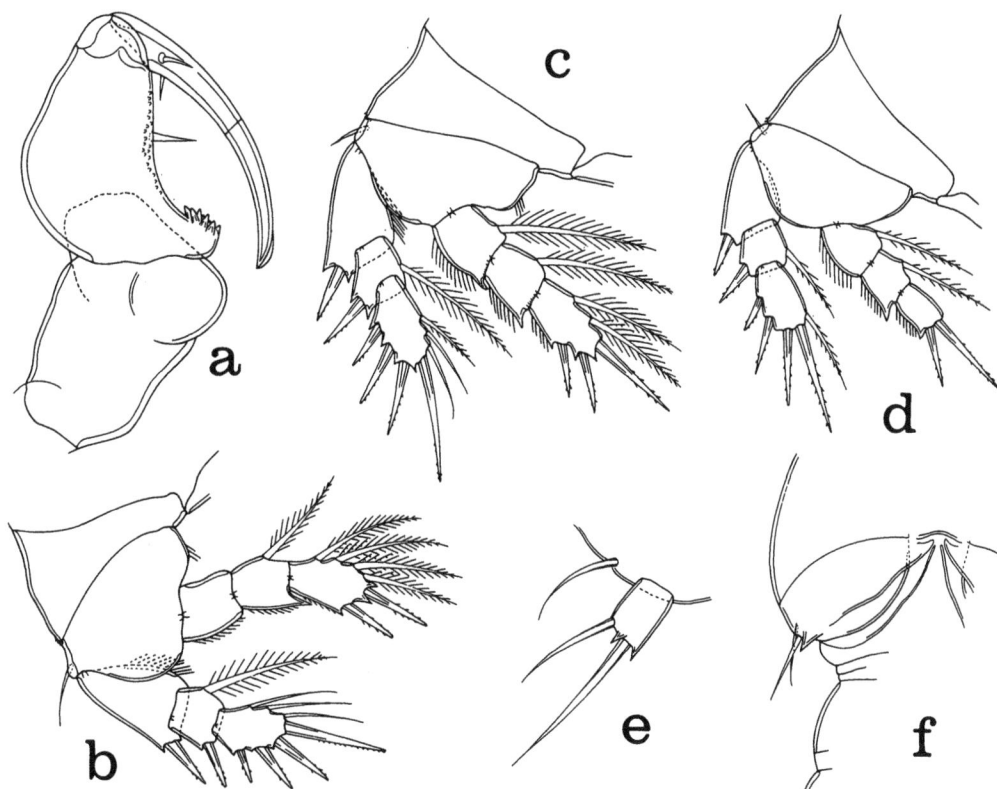

FIG. 29. *Lambanetes gemmulatus* gen. n., sp. n., male: a, maxilliped, antero-outer (C); b, leg 1, anterior (C); c, leg 2, anterior (C); d, leg 3, anterior (C); e, leg 5, lateral (G); f, leg 6, ventral (F).

Rostrum (Fig. 27e) weakly defined.

First antenna (Fig. 27f) 99 μm long. Lengths of seven segments: 12 (29 μm along anterior margin), 26, 10, 6, 8, 10, and 10 μm respectively. Formula for armature: 4, 13, 6, 2, 4 + 1 aesthete, 2 + 1 aesthete, and 7 + 1 aesthete. One seta on distal anterior margin of segment 2 spiniform. All setae smooth.

Second antenna (Fig. 27g) 88 μm long. Formula: 1, 1, 3, and 2 claws plus 5 setules. Claws recurved, larger claw 18 μm, smaller claw 11 μm.

Labrum (Fig. 27h) broad with median posteroventral process. Mandible (Fig. 27i) small, with convex side beyond indentation bearing broad minutely barbed spine followed by several small toothlike spines, and with concave side having short row of spinules. Lash short, spiniform, and barbed. Paragnath (Fig. 27h) a small lobe with hairs. First maxilla (Fig. 27j) with two setae. Second maxilla (Fig. 27k) 2-segmented. First segment unarmed. Second segment with surficial seta bearing only few lateral spinules. Inner seta short and smooth. Spiniform tip of segment with two very unequal setae on its base. Maxilliped (Fig. 27l) 3-segmented and short with usual lichomolgid armature. Second segment gibbous and bearing inner proximal digitiform lobe.

Ventral area between maxillipeds and first pair of legs (Fig. 27m) not protuberant and lacking median sclerite usually seen in lichomolgids in front of intercoxal plate of leg 1.

Legs 1–4 (Fig. 28a–d) with 3-segmented rami except for 2-segmented endopod of leg 4. Spine and setal formula as follows (Roman numerals representing spines, Arabic numerals indicating setae):

P_1	coxa	0-0	basis	1-0	exp	I-0;	I-1;	III, I, 3
					enp	0-0;	0-1;	I, 4
P_2	coxa	0-0	basis	1-0	exp	I-0;	I-1;	III, I, 3
					enp	0-1;	0-2;	I, II, 2
P_3	coxa	0-0	basis	1-0	exp	I-0;	I-1;	III, I, 2
					enp	0-1;	0-1;	II
P_4	coxa	0-0	basis	1-0	exp	I-0;	I-1;	III, I, 1
					enp	0-1;	II	

Legs 1–4 without inner coxal seta. Terminal segments of rami with reduced armature. Spines on exopods with posteriorly recurved tips. Setae on all four legs either smooth or at most with a few delicate hairs. Endopod of leg 1 with first segment lacking inner seta (Fig. 28a). Leg 4 exopod (Fig. 28d) 55 μm. First segment of endopod 16 × 18 μm with inner seta 10 μm. Second segment 20 × 12 μm with two terminal finely barbed spines 10 μm (outer) and 25 μm (inner).

Leg 5 (Fig. 28e) placed ventrolaterally, with unornamented free segment 35 × 16.5 μm (length including two small spiniform processes). Two terminal setae 25 μm and 35 μm. Dorsal seta adjacent to free segment about 20 μm. All three setae naked.

Setae representing leg 6 not visible on genital area.

Color in living specimens in transmitted light translucent gray, eye red.

Male.—Body (Fig. 28f, g) moderately elongate. Length of body 0.70 mm (0.63–0.75 mm) and greatest width 0.28 mm (0.26–0.30 mm), based on 10 specimens. Ratio of length to width of prosome 1.35:1. Ratio of length of prosome to that of urosome 1.30:1. Epimera of segments of legs 1–4 rounded (Fig. 28g), without budlike knobs.

Segment of leg 5 (Fig. 28h) short, dorsally 21 × 172 μm. Genital segment 104 × 177 μm, wider than long. Four postgenital segments from anterior to posterior 48 × 91, 49 × 82, 39 × 73, and 49 × 76 μm. Posteroventral margin of anal segment smooth.

Caudal ramus (Fig. 28h) resembling that of female but smaller, 34 × 18 μm.

Body surface smooth as in female.

Rostrum, first antenna, second antenna, labrum, mandible, paragnath, first maxilla, and second maxilla as in female. Maxilliped (Fig. 29a) large, 4-segmented, assuming that proximal part of claw represents fourth segment. First segment unarmed. Second segment with two unequal smooth setae. Proximal inner angle of segment protruding and bearing several dentiform spines; smaller spines along more distal inner side of segment. Small third segment unarmed. Claw 91 μm along its axis, with very narrow terminal lamella. Claw subdivided about midway and bearing proximally two short smooth setae.

Ventral area between maxillipeds and first pair of legs as in female.

Legs 1–4 with rami segmented as in female. Armature as in female except in one male the following:

P_1	enp	0-0;	0-1;	I, I, 4	(Fig. 29b)
P_2	enp	0-1;	0-1;	I, II, 2	(Fig. 29c)
P_3	exp	I-0;	I-1;	III, I, 1	(Fig. 29d)

Setae on these legs generally longer and more conspicuously feathered than in female.

Leg 5 (Fig. 29e) placed ventrolaterally, with small free segment 17 × 10 μm, its two terminal setae 33 μm and 44 μm. Dorsal seta about 25 μm.

Leg 6 (Fig. 29f) a posteroventral flap on genital segment bearing two naked setae 24 μm and 33 μm.

Spermatophore not seen.

Color in living specimens as in female.

Etymology.—The specific name *gemmulatus*, Latin meaning having buds, refers to the budlike knobs on the epimera of legs 2 and 3 in the female.

Remarks.—The male of *Lambanetes gemmulatus* may be distinguished from that of *L. stichodactylae* by its narrower genital segment, shorter claw on the maxilliped, various differences in the spine and setal armature of legs 1–4, and the relative length of the free segment of leg 5. The chief points of distinction between the two species are shown in Table 3.

Genus *Metaxymolgus* Humes and Stock, 1973

Metaxymolgus cuspis (Humes, 1964)
= *Lichomolgus cuspis* Humes, 1964

Host: *Heteractis magnifica* (Quoy and Gaimard) [= *Radianthus ritteri* (Kwietniewski)].

Site: In washings.

Locality: Region of Nosy Bé, northwestern Madagascar (Humes, 1964).

Notes: Length of ♀ 1.53 mm, ♂ 1.30 mm.

Host: *Stichodactyla gigantea* (Forskål) [= *Stoichactis giganteum* (Forskål)].

Site: In washings.

Localities: Region of Nosy Bé, northwestern Madagascar (Humes, 1964); near Nouméa, New Caledonia, new records as follows: 75 ♀♀, 31 ♂♂, 15 copepodids from one host, in 0.5 m, eastern edge of Isle Maître, near Nouméa, 22°20′35″S, 166°25′10″E, 8 June 1971; 42 ♀♀, 16 ♂♂, 2 copepodids from one host, intertidal, in sand, Ricaudy Reef, near Nouméa, 22°19′00″S, 166°26′44″E, 21 July 1977; 10 ♀♀, 12 ♂♂. 1 copepodid, from one host, in 0.5 m, Ricaudy Reef, near Nouméa, 5 June 1971; 18 ♀♀, 17 ♂♂, 6 copepodids from one host, in 25 cm, Ricaudy Reef, near Nouméa, 19 June 1971.

Citation: Bouligand (1966).

TABLE 3. Principal distinguishing features of the males
of the two species of *Lambanetes*

	L. stichodactylae	*L. gemmulatus*
Genital segment	156 × 335 μm	104 × 177 μm
Length of longest terminal seta on caudal ramus	about four times greatest length of ramus	about two times greatest length of ramus
First antenna	189 μm, one minutely barbed seta on third segment	99 μm, all setae smooth
Second maxilla	first segment gibbous	first segment not gibbous
Claw of maxilliped	192 μm	91 μm
First segment of P₁ Enp	0-1	0-0
Second segment of P₂ + P₃ Enp	0-2	0-1
Third segment of P₄ Exp	III, I, 2	III, I, 1
Free segment of P₅	65 × 13 μm	17 × 10 μm

Metaxymolgus myorae (Greenwood, 1971)
= *Lichomolgus myorae* Greenwood, 1971

Host: *Stichodactyla haddoni* (Saville-Kent) [= *Stoichactis haddoni* (Saville-Kent)].
Site: Upper and outer surface of body column.
Locality: Moreton Bay, Queensland, Australia (Greenwood, 1971).
Notes: Length of ♀ 1.50 mm, ♂ 1.48 mm.

Metaxymolgus pertinax sp. n.
Figs. 30a–g, 31a–i, 32a–g, 33a–e

Type material.—16 ♀♀, 18 ♂♂, and 3 copepodids from the actiniarian *Tealia coriacea* (Cuvier), Shell Beach, north of Bodega Bay, Sonoma County, California, 6 October 1979, S. Lönning and W. Vader coll. Holotype ♀, allotype, and 27 paratypes (12 ♀♀, 15 ♂♂) deposited in the National Museum of Natural History, Smithsonian Institution, Washington, D.C.; the remaining paratypes (dissected) and the copepodids in the collection of the author.

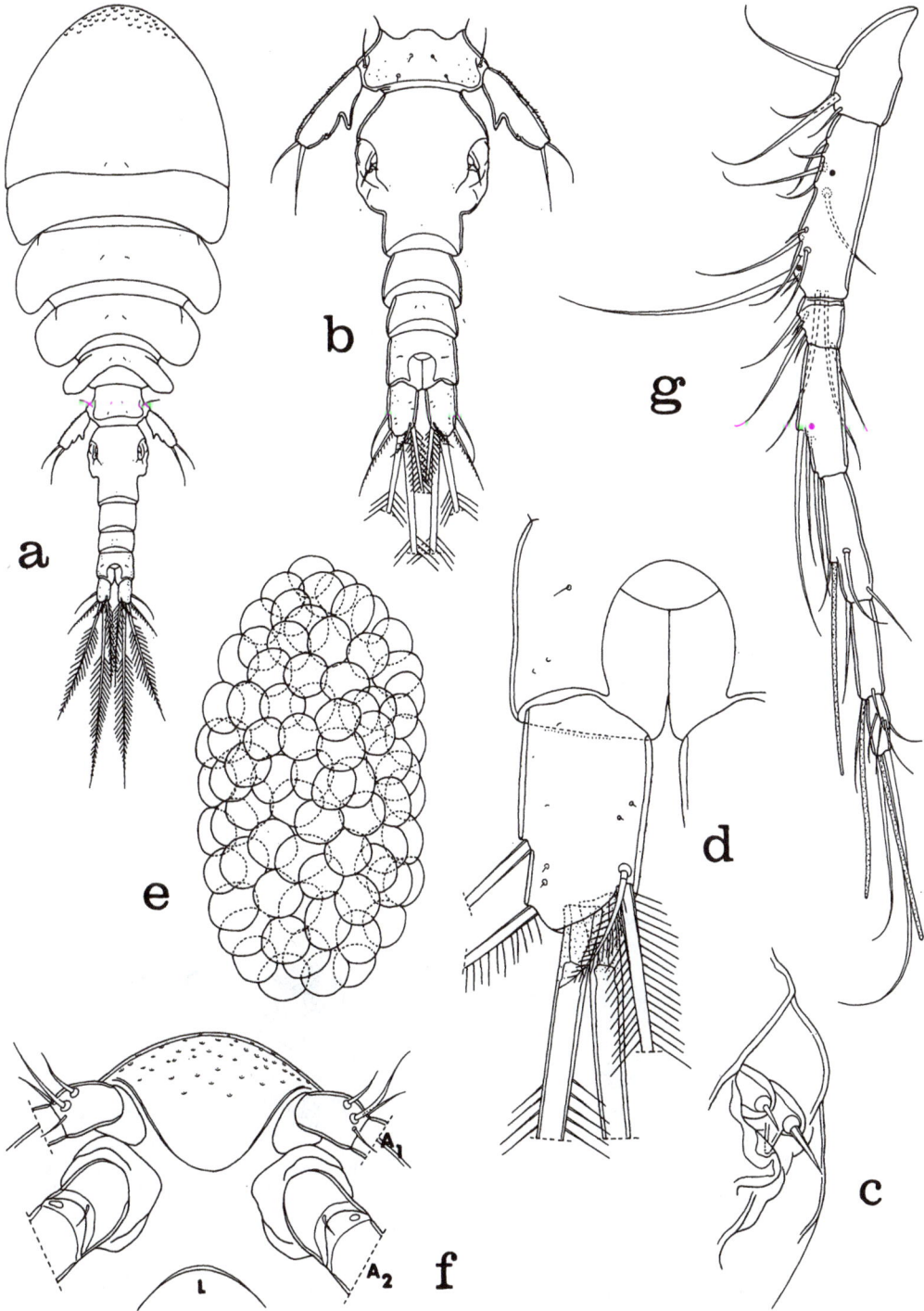

FIG. 30. *Metaxymolgus pertinax* sp. n., female: a, dorsal (A); b, urosome, dorsal (B); c, genital area, dorsal (C); d, caudal ramus, dorsal (C); e, egg sac, ventral (B); f, rostrum, ventral (E); g, first antenna, with dots indicating positions of aesthetes added in male, ventral (D).

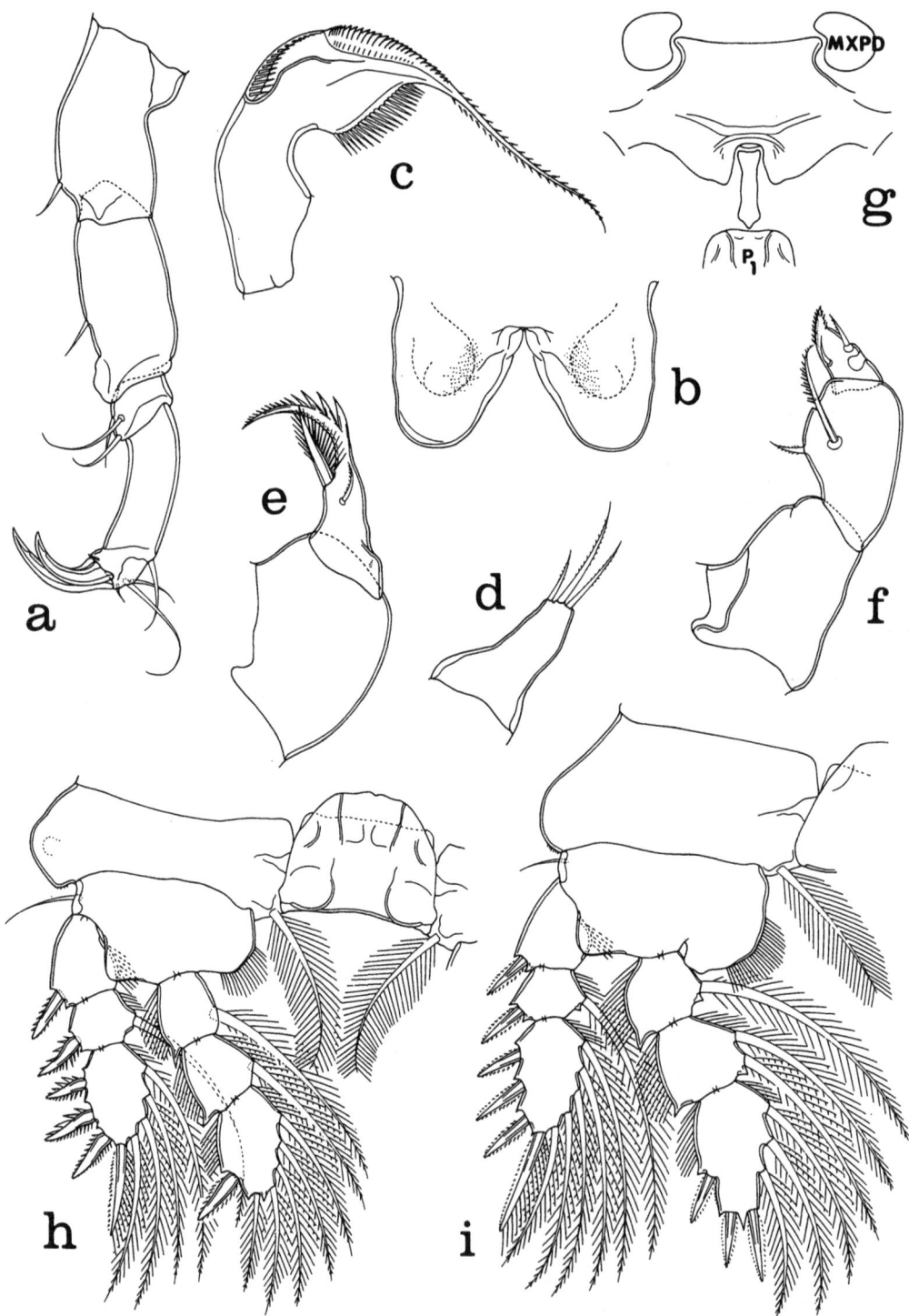

FIG. 31. *Metaxymolgus pertinax* sp. n., female: a, second antenna, posterior (D); b, labrum, with paragnaths indicated by broken lines, ventral (F); c, mandible, posterior (C); d, first maxilla, posterior (C); e, second maxilla, posterior (F); f, maxilliped, postero-inner (F); g, area between maxillipeds and first pair of legs, ventral (E); h, leg 1 and intercoxal plate, anterior (D); i, leg 2, anterior (D).

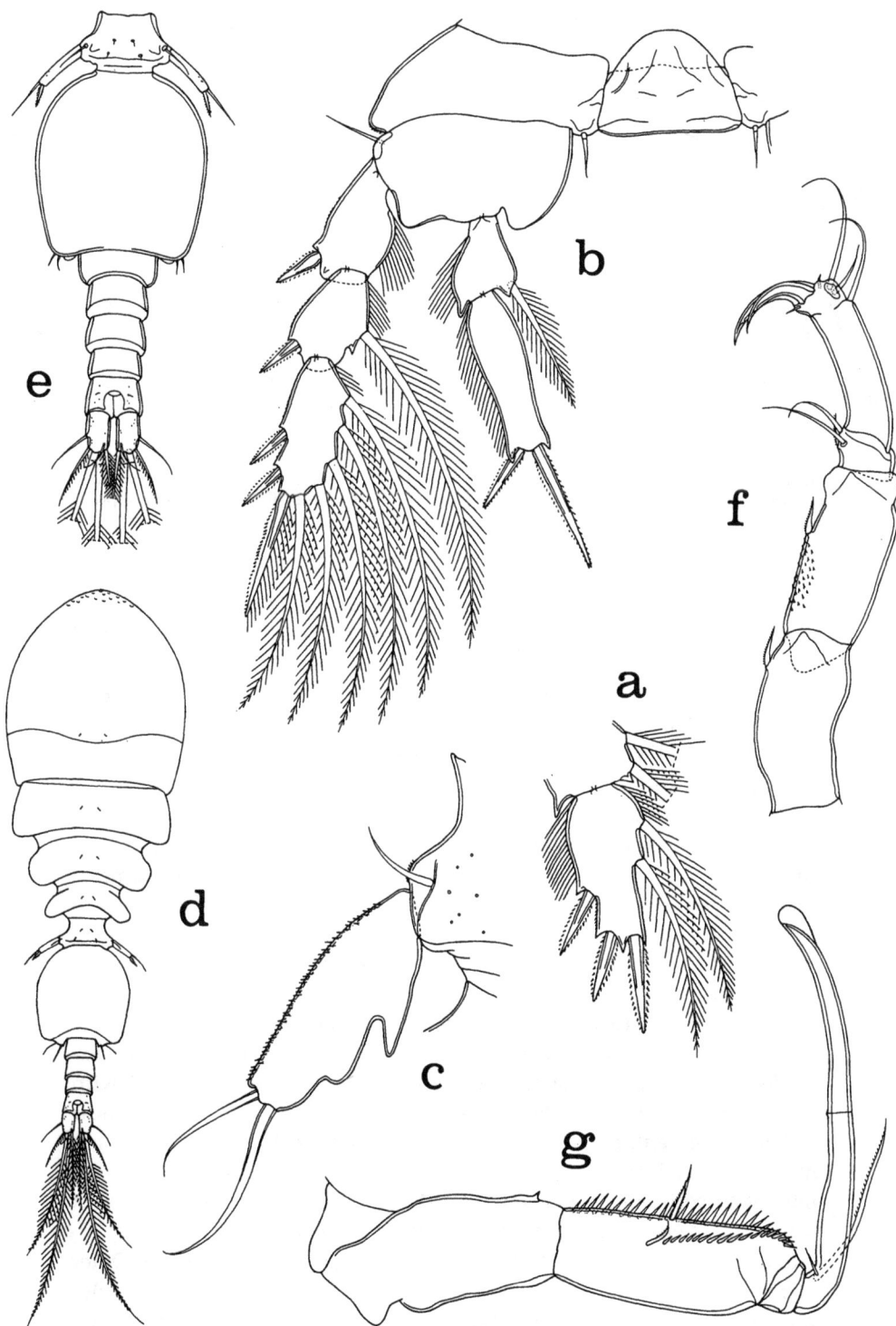

FIG. 32. *Metaxymolgus pertinax* sp. n. Female: a, third segment of endopod of leg 3, anterior (D); b, leg 4 and interxocal plate, anterior (D); c, leg 5, dorsal (F). Male: d, dorsal (E); e, urosome, dorsal (B); f, second antenna, posterior (D); g, maxilliped, outer (D).

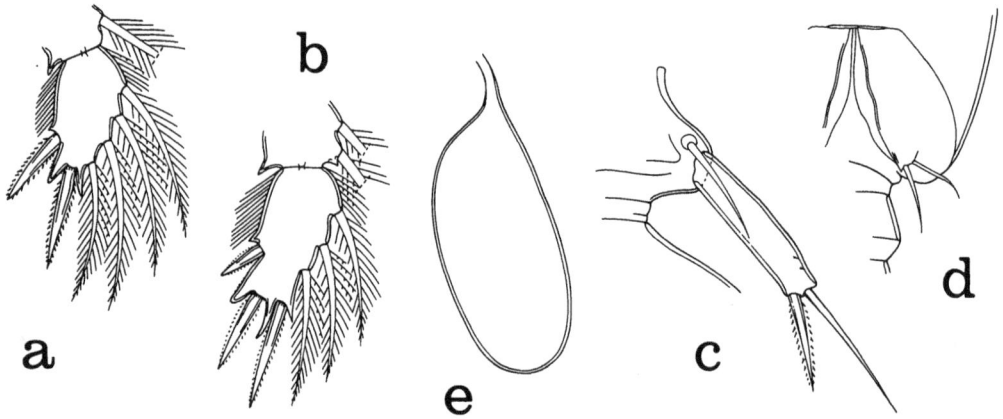

FIG. 33. *Metaxymolgus pertinax* sp. n., male: a, third segment of endopod of leg 1, anterior (D); b, third segment of endopod of leg 2, anterior (D); c, leg 5, dorsal (F); d, leg 6, ventral (E); e, spermatophore, extruded from male in lactic acid (E).

Other specimens.—From *Tealia crassicornis* (Müller): 5 ♀♀, 7 ♂♂, Shell Beach, Bodega Bay, Sonoma County, California, 1 November 1979, S. Lönning and W. Vader coll.

From *Tealia coriacea*: 4 ♀♀, in 0.6 m, Yaquina Bay, Lincoln County, Oregon, 1 August 1974, J. Ratliff coll.: 22 ♀♀, 4 ♂♂, Shell Beach, Bodega Bay, California, 15 November 1979, S. Lönning and W. Vader coll.

Note.—This species is referred to by Lönning and Vader in manuscript as "*Metaxymolgus* sp. D".

Female.—Body (Fig. 30a) moderately slender. Length 1.52 mm (1.35–1.68 mm) and greatest width 0.65 mm (0.50–0.79 mm), based on 10 specimens. First pedigerous segment separated dorsally from cephalosome by transverse suture. Epimera of segments of leg 1–4 rounded. Ratio of length to width of prosome 1.62:1. Ratio of length of prosome to that of urosome 1.73:1.

Segment of leg 5 (Fig. 30b) 100 × 198 μm. Genital segment in dorsal view 234 × 211 μm, slightly longer than wide, anterior two-thirds a little expanded with irregular margins, posterior third abruptly narrowed with parallel sides (width here 125 μm). Genital areas situated dorsolaterally near middle of segment. Each area (Fig. 30c) with two small naked setae 11 μm and 24 μm and a small spiniform process. Three postgenital segments from anterior to posterior 78 × 114, 62 × 104, and 78 × 109 μm. Anal segment with posteroventral row of minute spinules on both sides.

Caudal ramus (Fig. 30d) 78 × 34 μm, longer than wide, ratio 1.66:1. Outer lateral seta 90 μm and naked, dorsal seta about 33 μm and delicately plumose, outermost terminal seta 120 μm with hairs along inner side, and innermost terminal seta 224 μm and feathered. Two median terminal setae 418 μm (outer) and 570 μm (inner), both with lateral spinules and inserted between dorsal flange (smooth) and ventral flange (with minute spinules).

Body surface with a few hairs (sensilla) and refractile points, the latter numerous on rostral area and front of head (Fig. 30a).

Egg sac (Fig. 30e) elongate oval, 653 (418–770) × 337 (242–385) μm, based on five specimens in lactic acid. Eggs numerous, 68–78 μm in diameter and slightly irregular in shape.

Rostrum (Fig. 30f) rounded posteroventrally.

First antenna (Fig. 30g) 523 μm long. Lengths of seven segments: 42 (78 μm along anterior margin), 122, 39, 88, 88, 68, and 40 μm respectively. Formula for armature: 1, 13, 6, 3, 4 + 1 aesthete, 2 + 1 aesthete, and 7 + 1 aesthete. All setae naked.

Second antenna (Fig. 31a) 392 μm long and 4-segmented, with armature 1, 1, 3, and 2 claws plus 5 setules. Fourth segment 135 μm along outer side, 87 μm along inner side, and 38 μm wide. Two terminal claws unequal, longer claw 83 μm along its axis, shorter claw 62 μm. All setae naked.

Labrum (Fig. 31b) with two rounded posteroventral lobes. Mandible (Fig. 31c) slender, having on concave side beyond constriction of base a row of spinules and on convex side a scalelike area bearing row of spinules followed by serrate fringe. Lash long and barbed. Paragnath (Fig. 31b) a small hairy lobe. First maxilla (Fig. 31d) with three finely barbed setae. Second maxilla (Fig. 31e) and maxilliped (Fig. 31f) resembling in general respects those of other species of *Metaxymolgus*, for example, *Metaxymolgus praelongipes* Humes, 1975, though differing in minor details.

Ventral area between maxillipeds and first pair of legs (Fig. 28g) not protuberant.

Legs 1–4 (Figs. 31h, i, 32a, b) segmented and armed as in other species of genus. Coxa of leg 1 with small outer prominence on posterior surface (Fig. 31h). Barbs on exopod spines of leg 1 conspicuously large but smaller in succeeding legs. Leg 4 (Fig. 32b) with inner coxal seta 29 μm, naked, with faint proximal joint (this seta in preceding legs long and plumose). Inner margin of basis smooth (haired in preceding legs). Exopod 244 μm long. First segment of endopod 55 μm long without spiniform processes (62 μm with these processes) and 47 μm wide, its inner plumose seta 104 μm. Second segment 113 μm long without spiniform processes (118 μm with these processes) and 39 μm in greatest width; two terminal barbed spines 51 μm (outer) and 91 μm (inner), both with minutely bifurcate tips.

Leg 5 (Fig. 32c) elongate, with free segment 125 μm long, its inner margin having two processes, proximal process large and distally directed (width of segment here 59 μm), distal process smaller and rounded (width of segment here 46 μm). Ratio of length to greatest width of free segment 2.12:1. Two terminal smooth setae 65 μm and 88 μm. Free segment ornamented with small spines along outer edge. Dorsal seta near insertion of free segment approximately 49 μm. Row of minute spines on outer corner of segment of leg 5 near insertion of free segment.

Leg 6 represented by two setae on genital area (Fig. 30c).

Male.—Body (Fig. 32d) a little more slender than in female. Length 1.49

mm (1.39–1.57 mm) and greatest width 0.53 mm (0.48–0.56 mm), based on 10 specimens. Ratio of length to width of prosome 1.74:1. Ratio of length of prosome to that of urosome 1.53:1.

Segment of leg 5 (Fig. 32e) 57 × 151 μm. Genital segment 360 × 273 μm, longer than wide. Four postgenital segments from anterior to posterior 52 × 104, 49 × 95, 39 × 89, and 57 × 94 μm.

Caudal ramus similar to that of female but smaller, 62 × 39 μm, ratio 1.59:1.

Rostrum like that of female. First antenna resembling that of female but three long aesthetes added, two on second segment and one on fourth segment (at points indicated by dots in Figure 32g). Formula thus: 4, 13 + 2 aesthetes, 6, 3 + 1 aesthete, 4 + 1 aesthete, 2 + 1 aesthete, and 7 + 1 aesthete. All setae smooth as in female. Second antenna (Fig. 32f) similar to that of female but showing sexual dimorphism in seta on first and second segments and one seta on third segment, all three of which are finely pectinate along one side, and in second segment with inner surface bearing small spines.

Labrum, mandible, paragnath, first maxilla, and second maxilla like those of female. Maxilliped (Fig. 32g) slender and 4-segmented (considering proximal half of claw to represent fourth segment). First segment with minute inner distal thornlike process. Second segment with two barbed setae and two rows of spines. Small third segment unarmed. Claw 280 μm long including terminal lamella, with subdivision midway along its length, and bearing two extremely unequal proximal setae, longer seta barbed unilaterally, smaller seta naked.

Ventral area between maxillipeds and first pair of legs as in female.

Legs 1–4 segmented as in female with same spine and setal formula as in that sex except for third segment of endopod of leg 1 (Fig. 33a) where formula is I, I, 4 instead of I, 5 as in female. Additional sexual dimorphism seen in greater length of terminal spiniform processes on third segment of endopod of leg 1 and of leg 2 (Fig. 33b). Otherwise legs 1–4 resembling those of female.

Leg 5 (Fig. 33c) with free segment slender, 79 × 16 μm, ratio 4.94:1. Inner terminal element a barbed spine 39 μm, outer terminal element a smooth seta 62 μm. Free segment ornamented with only two or three very small distal spines. Outer corner of segment of leg 5 smooth.

Leg 6 (Fig. 33d) the usual posteroventral flap on genital segment, bearing two naked setae approximately 60 μm long.

Spermatophore (Fig. 33e), extruded from male, about 260 × 117 μm, not including neck.

Color and habitat notes.—Lönning and Vader (in manuscript) describe the color of *Metaxymolgus pertinax* (their sp. D) as follows: Specimens on *Tealia coriacea* "are usually off-white with a faint reddish hue. They live mainly on and among the tentacles and on the mouthfield, and do not emerge when the sea anemone contracts." Those on *Tealia crassicornis* "are often quite vividly colored, with dark green or dark red 'digestive cross' and egg sacs, similar to but even darker than the red and green blotches on the column of the sea anemones. The copepods live on this host mainly on the column and are less often seen on the tentacles."

Etymology.—The specific name *pertinax*, Latin meaning holding fast, alludes to the habit of this copepod of clinging to the host.

Remarks.—The structure of the fifth legs in the female distinguishes *Metaxymolgus pertinax* from all its congeners. No other species has two processes on the inner side of the free segment of leg 5.

Metaxymolgus confinis sp. n.
Figs. 34a–k, 35a–i

Type material.—18 ♀♀, 8 ♂♂, 1 copepodid from the actiniarian *Anthopleura elegantissima* (Brandt), Shell Beach, north of Salmon Creek, Sonoma County, California, 21 October 1979, S. Lönning and W. Vader coll. Holotype ♀, allotype, 24 paratypes (17 ♀♀, 7 ♂♂), and the copepodid deposited in the National Museum of Natural History, Smithsonian Institution, Washington, D.C.

Other specimens.—From *Anthopleura elegantissima*: 11 ♀♀, 7♂♂, 1 copepodid, Doran Rocks, Bodega Bay, California, 11 August 1979, S. Lönning and W. Vader coll.; 19 ♀♀, 5 ♂♂, Bodega Bay, California, 11 December 1979, S. Lönning and W. Vader coll.; 10 ♀♀, 2 ♂♂, 4 copepodids, south jetty, Bodega Bay, 11 July 1979, G. Ruiz coll.; 7 ♀♀, 2 ♂♂, 6 copepodids, north jetty, Bodega Bay, 12 July 1979, G. Ruiz coll.; 13 ♀♀, 4 ♂♂, 3 copepodids, exposed area, Bodega Bay, 12 July 1979, G. Ruiz coll.; 21 ♀♀, 2 ♂♂, 2 copepodids, Coal Oil Point, Goleta, Santa Barbara, California, 17 December 1979, S. Lönning and W. Vader coll.; 21 ♀♀, 2 ♂♂, 2 copepodids, Bird Rock, Hopkins Marine Station, Pacific Grove, California, 19 December 1979, S. Lönning and W. Vader coll.; 6 ♀♀, 1 ♂, 2 copepodids, Hopkins Marine Station, Pacific Grove, California, 15 December 1979, S. Lönning and W. Vader coll.

From *Anthopleura xanthogrammica* (Brandt): 11 ♀♀, 12 ♂♂, Bodega Bay, California, 12 December 1979, S. Lönning and W. Vader coll.; 58 ♀♀, 20 ♂♂, 5 copepodids, north jetty, Bodega Bay, 13 July 1979, G. Ruiz coll.; 31 ♀♀, 10 ♂♂, 9 copepodids, south jetty, Bodega Bay, 12 July 1979, G. Ruiz coll.

From *Tealia piscivora* Sebens and Laakso: 9 ♀♀, 6 ♂♂, and 1 copepodid in 4 m, between the jetties, channel of Bodega Harbor, Bodega Bay, California, 9 June 1980, Gregory Ruiz coll.

Note.—This species is referred to by Lönning and Vader in manuscript as "*Metaxymolgus* sp. B".

Female.—Body (Fig. 34a) resembling in general form that of *Metaxymolgus pertinax*, though prosome a little more pointed anteriorly than in that species. Length 1.54 mm (1.29–0.66 mm) and greatest width 0.64 mm (0.58–0.72 mm), based on 10 specimens. Ratio of length to width of prosome 1.53:1. Ratio of length of prosome to that of urosome 1.55:1.

Segment of leg 5 (Fig. 34b) 99 × 203 μm. Genital segment in dorsal view 234 × 234 μm, as long as wide, insected in posterior fourth. Genital areas located dorsolaterally anterior to middle of segment. Each area resembling that of *M. pertinax*, the two setae 7 μm and 11 μm. Three postgenital seg-

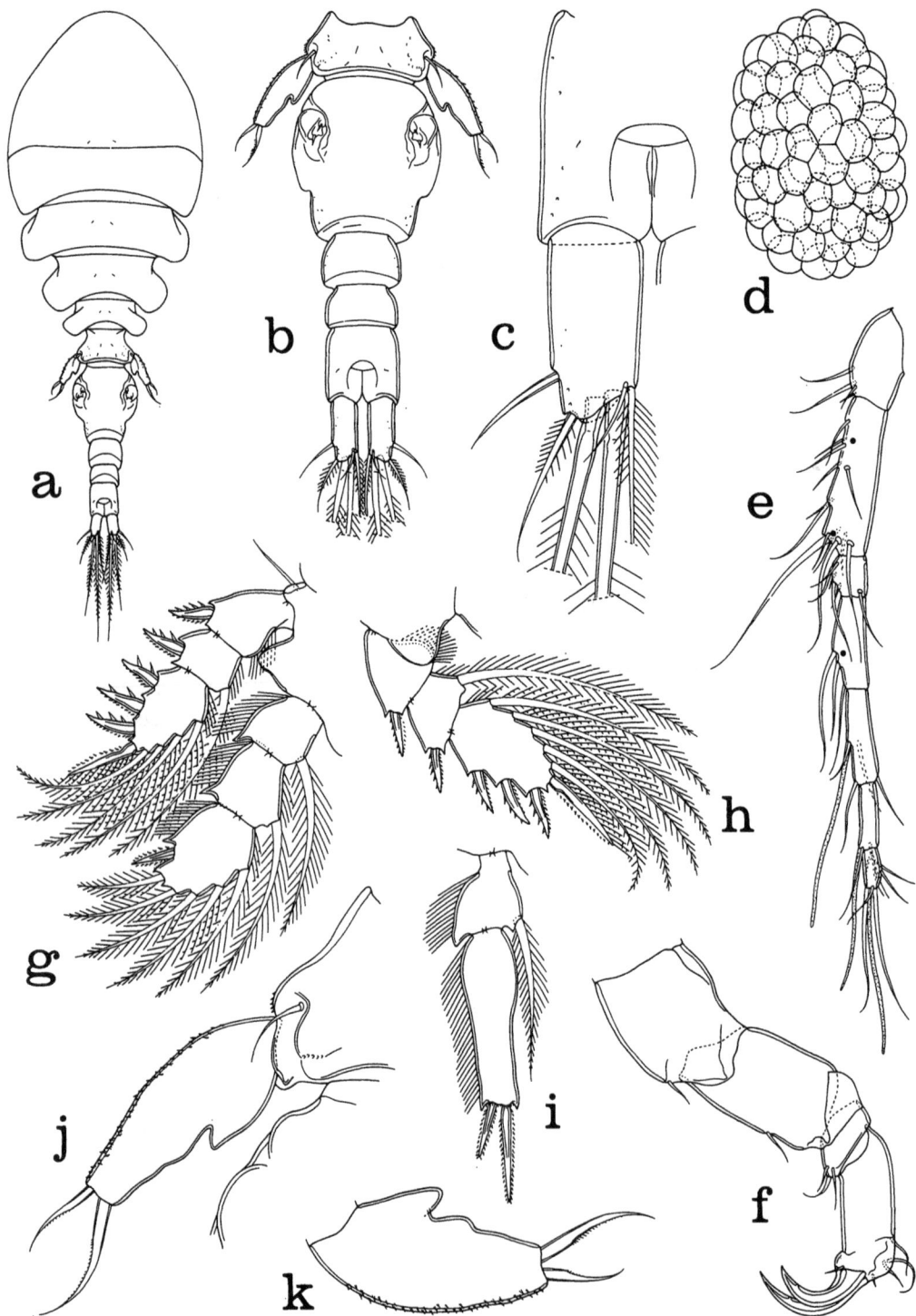

FIG. 34. *Metaxymolgus confinis* sp. n., female: a, dorsal (A); b, urosome, dorsal (B); c, caudal ramus, dorsal (F); d, egg sac, lateral (B); e, first antenna, dorsal (D); f, second antenna, posterior (D); g, rami of leg 1, anterior (D); h, exopod of leg 2, anterior (D); i, endopod of leg 4, anterior (D); j, leg 5, dorsal (F); k, free segment of leg 5, dorsal (F).

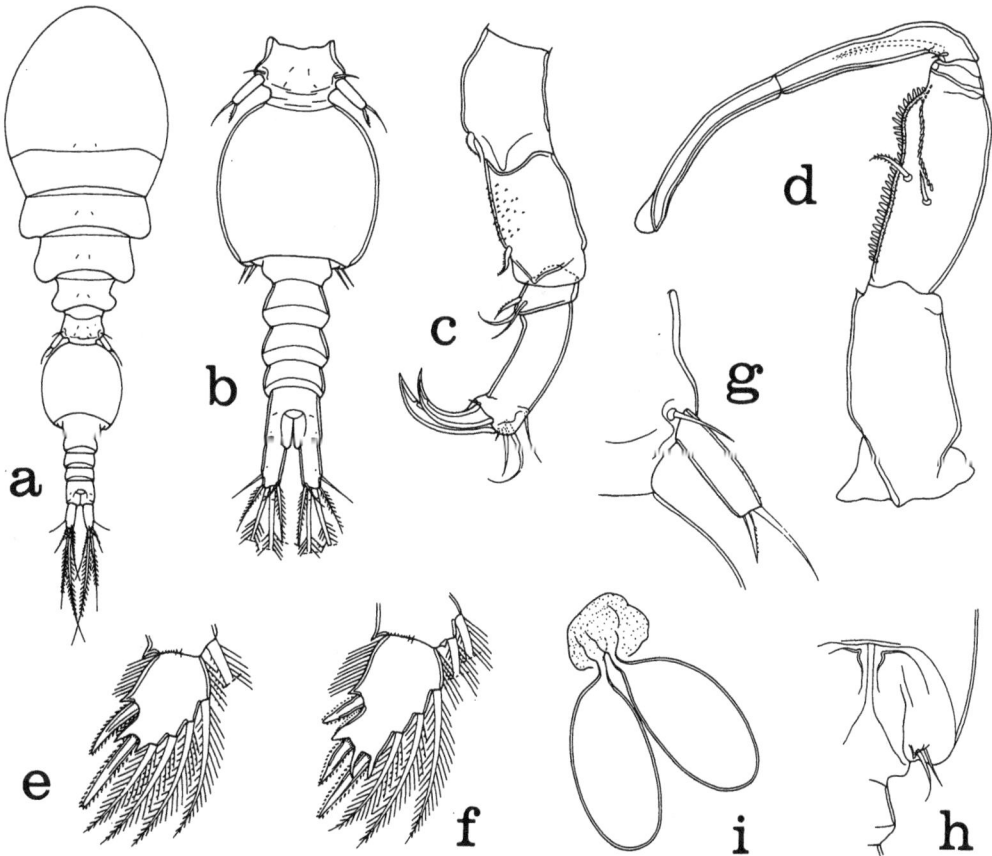

FIG. 35. *Metaxymolgus confinis* sp. n., male: a, dorsal (A); b, urosome, dorsal (B); c, second antenna, posterior (D); d, maxilliped, outer (D); e, third segment of endopod of leg 1, anterior (D); f, third segment of endopod of leg 2, anterior (D); g, leg 5, dorsal (F); h, leg 6, ventral (E); i, spermatophores, attached to female, dorsolateral (B).

ments from anterior to posterior 78 × 128, 57 × 114, and 109 × 114 μm. Anal segment with posteroventral row of minute spinules on both sides.

Caudal ramus (Fig. 34c) 95 × 44 μm, ratio 2.16:1. Outer lateral seta 60 μm and dorsal seta 27 μm, both smooth. Outermost terminal seta 73 μm and delicately plumose proximally, and innermost terminal seta 120 μm and feathered. Two median terminal setae 255 μm (outer) and 350 μm (inner), both with strong lateral spinules and inserted between small dorsal flange (smooth) and ventral flange (also smooth).

Body surface with very few hairs (sensilla) and refractile points (Fig. 34a).

Egg sac (Fig. 34d) oval, 396 × 280 μm, eggs numerous and 49–68 μm in diameter.

Rostrum as in *M. pertinax*.

First antenna (Fig. 34e) 396 μm long. Lengths of seven segments: 26 (61 μm along anterior margin), 104, 28, 65, 62, 47, and 29 μm respectively. Formula for armature as in *M. pertinax*. All setae naked.

Second antenna (Fig. 34f) 308 μm long, segmented and armed as in *M. pertinax*. Fourth segment 109 μm along outer side, 60 μm along inner side, and 41 μm wide, relatively shorter than in *M. pertinax*. Two terminal claws unequal, longer claw 78 μm, shorter claw 62 μm. All setae naked.

Labrum, mandible, paragnath, first maxilla, second maxilla, maxilliped, and ventral area between maxillipeds and first pair of legs as in *M. pertinax*.

Legs 1–4 segmented and armed as in *M. pertinax*. Leg 1 with barbs on exopod spines larger and coarser (Fig. 34g) than in *M. pertinax*. Leg 2 with barbs on exopod spines a little larger (Fig. 34h) than in *M. pertinax*. Leg 3 entirely as in *M. pertinax*. Leg 4 with endopod (Fig. 34i) differing only slightly from *M. pertinax*. First segment of endopod 52 μm long without spiniform processes (58 μm with these processes) and 49 μm wide, its inner plumose seta 117 μm. Second segment 122 μm without spiniform processes (130 μm with these processes) and 35 μm in greatest width; two terminal barbed spines 40 μm (outer) and 70 μm (inner).

Leg 5 (Fig. 34j) elongate, with free segment 127 μm long, its inner margin with pointed proximal process (width of segment here 52 μm) and slight distal swelling (width of segment here 47 μm). Ratio of length to greatest width of free segment 2.44:1. Two terminal setae 42 μm and 64 μm. Dorsal seta near insertion of free segment about 36 μm. Free segment ornamented with small spines along outer edge. One female with free segment 120 × 64 μm (Fig. 34k). Few minute spinules on outer corner of segment of leg 5 near insertion of free segment.

Leg 6 represented by two setae on genital area.

Male.—Body (Fig. 35a) slender. Length 1.39 mm (1.32–1.50 mm) and greatest width 0.45 mm (0.41–0.53 mm), based on 10 specimens. Ratio of length to width of prosome 1.85:1. Ratio of length of prosome to that of urosome 1.46:1.

Segment of leg 5 (Fig. 35b) 78 × 122 μm. Genital segment 237 × 224 μm, only slightly longer than wide. Four postgenital segments from anterior to posterior 52 × 88, 55 × 85, 36 × 75, and 68 × 78 μm.

Caudal ramus 68 × 31 μm, ratio 2.19:1, resembling that of female.

Rostrum like that of *M. pertinax*. First antenna resembling that of female but three long aesthetes added (at positions indicated by dots in Figure 34e), so that formula is same as for male of *M. pertinax*. Second antenna (Fig. 35c) armed and ornamented as in male of *M. pertinax*.

Labrum, mandible, paragnath, first maxilla, and second maxilla as in *M. pertinax*. Maxilliped (Fig. 35d) resembling that of *M. pertinax*, claw 252 μm long.

Ventral area between maxillipeds and first pair of legs as in *M. pertinax*.

Legs 1–4 resembling those of male of *M. pertinax*, with similar sexual dimorphism. Third endopod segment of leg 1 (Fig. 35e) and third endopod segment of leg 2 (Fig. 35f) differing only slightly from male of *M. pertinax*.

Leg 5 (Fig. 35g) with free segment 55 × 14 μm, ratio 3.93:1. Inner terminal element 21 μm, outer terminal element 41 μm. Dorsal seta 27 μm. Free segment ornamented with few minute outer spinules. Outer corner of segment of leg 5 smooth.

Leg 6 (Fig. 35h) the usual posteroventral flap on genital segment, bearing two naked setae 40 μm (stout) and 36 μm (more slender).

Spermatophore (Fig. 35i) 231 × 117 μm without neck.

Color and habitat notes.—*Metaxymolgus confinis* (Lönning and Vader's sp. B) on *Anthopleura elegantissima* is "quite variable in color, but the color shades are always subdued; the egg sacs are white or pinkish. The copepods spend most of their time on and among the tentacles and the upper part of the column. When the sea anemone contracts, the copepods usually disappear from view." Specimens of *M. confinis* on *Anthopleura xanthogrammica* are "generally of a very pale color, making them quite easy to pick out on the green column and tentacles of their host. They mainly live on and among the tentacles " (Lönning and Vader, in manuscript).

Etymology.—The specific name *confinis*, Latin meaning adjacent or neighboring, alludes to the close relationship of this species with *Metaxymolgus pertinax*.

Remarks.—*Metaxymolgus confinis* is taxonomically close to *Metaxymolgus pertinax* but the new species differs from *M. pertinax* in several respects, the most obvious of these being the relative length of the fourth segment of the second antenna, and the form of the free segment of leg 5 (see Table 4).

Metaxymolgus turmalis sp. n.
Figs. 36a–g, 37a–e

Type material.—5 ♀♀, 7 ♂♂, 19 copepodids from the actiniarian *Anthopleura artemisia* (Pickering, in Dana), south jetty, Bodega Bay, California, 11 July 1979. G. Ruiz collector. Holotype ♀, allotype, and 8 paratypes (3 ♀♀, 5 ♂♂) deposited in the National Museum of Natural History, Smithsonian Institution, Washington, D.C.; the remaining paratypes (dissected) and the copepodids in the collection of the author.

Other specimens.—4 ♀♀, 3♂♂, 8 copepodids from *Anthopleura artemisia*, north jetty, Bodega Bay, California, July 1979. G. Ruiz collector.

Note.—This species is referred to by Lönning and Vader in manuscript as "*Metaxymolgus* sp. A".

Female.—Body (Fig. 36a) slender. Length 1.40 mm (1.34–1.44 mm) and greatest width 0.56 mm (0.54–0.57 mm), based on 5 specimens. Epimera of segments of legs 1–4 rounded. Segment of leg 1 separated dorsally from head by transverse furrow. Ratio of length to width of prosome 1.74:1. Ratio of length of prosome to that of urosome 1.87:1.

Segment of leg 5 (Fig. 36b) 90 × 164 μm. Genital segment 205 × 180 μm, insected in posterior fourth. Genital areas situated dorsolaterally just anterior to middle of segment. Each area as in *Metaxymolgus pertinax*, with two setae 10 μm and 18 μm. Three postgenital segments from anterior to posterior 70 × 96, 49 × 86, and 55 × 94 μm. Anal segment with posteroventral row of minute spinules on both sides.

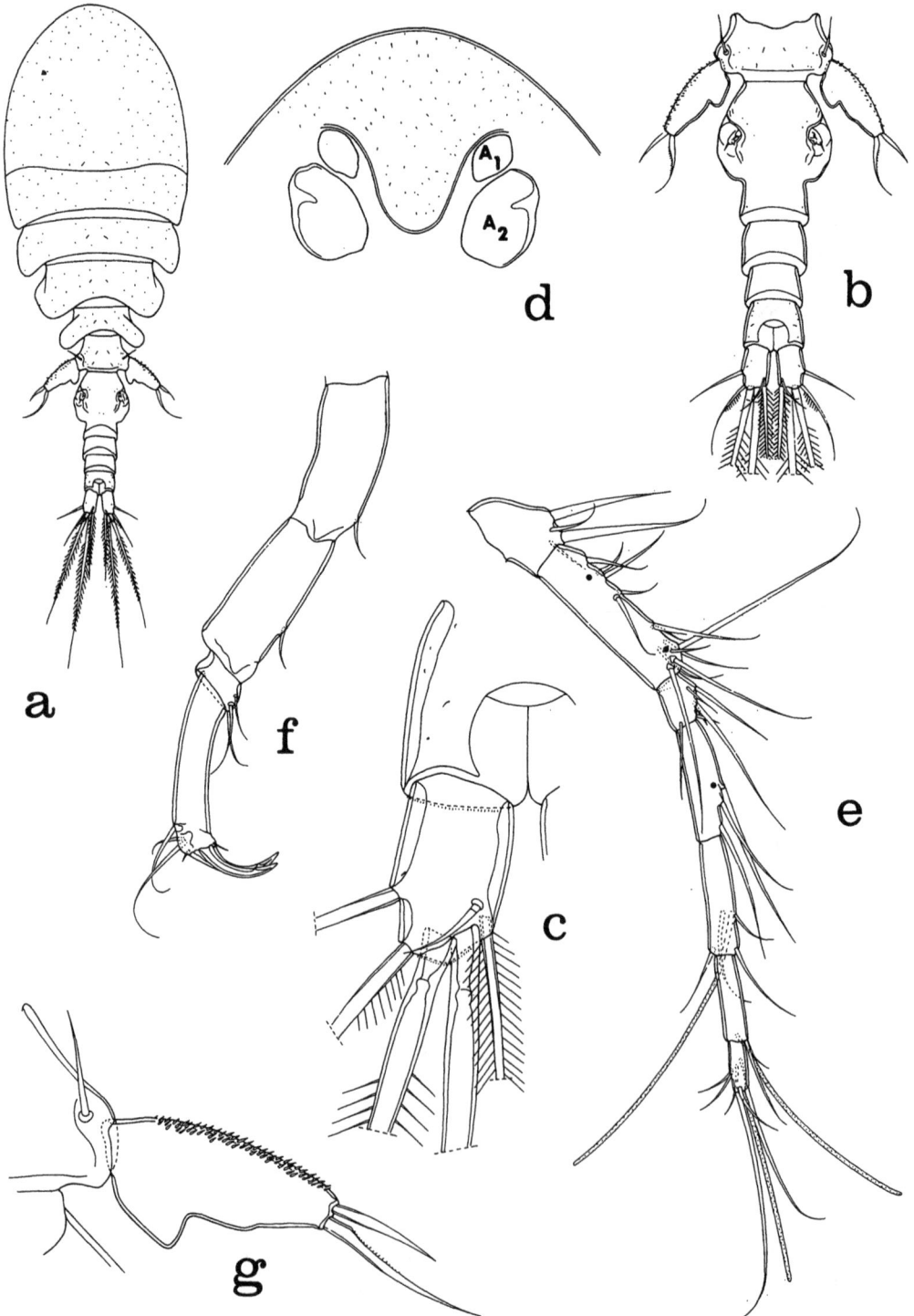

FIG. 36. *Metaxymolgus turmalis* sp. n., female: a, dorsal (A); b, urosome, dorsal (B); c, caudal ramus, dorsal (C); d, rostrum, ventral (E); e, first antenna, dorsal (D); f, second antenna, posterior (D); g, leg 5, dorsal (F).

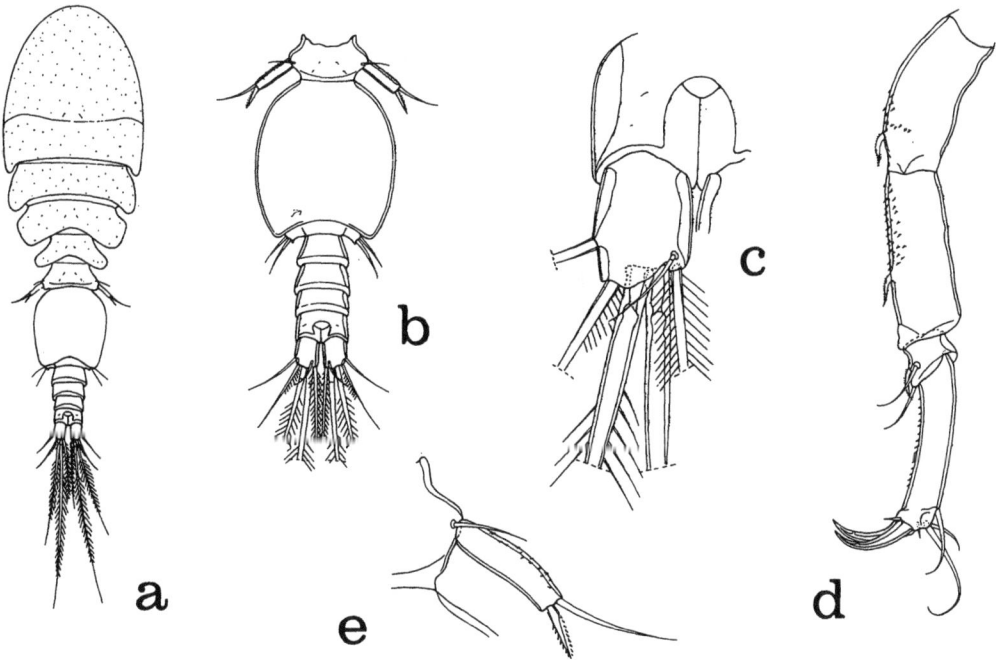

FIG. 37. *Metaxymolgus turmalis* sp. n., male: a, dorsal (A); b, urosome, dorsal (E); c, caudal ramus, dorsal (C); d, second antenna, posterior (D); e, leg 5, dorsal (F).

Caudal ramus (Fig. 36c) 65 × 43 μm, only a little longer than wide, ratio 1.51:1. Outer lateral seta 96 μm and dorsal seta about 33 μm, both smooth. Outermost terminal seta 127 μm, with proximal inner hairs. Innermost terminal seta 192 μm and feathered. Two long median terminal setae 340 μm (outer) and 473 μm (inner) both with lateral spinules and inserted between small dorsal flange (smooth) and ventral flange (with marginal spinules).

Body surface with few hairs (sensilla) but many refractile points over dorsal surface of prosome (Fig. 36a).

Egg sac as in *M. pertinax*, 540 × 265 μm, eggs 42–55 μm in diameter, reaching just beyond caudal rami.

Rostrum (Fig. 36d) with rounded posteroventral margin.

First antenna (Fig. 36e) segmented and armed as in *M. pertinax*. Lengths of seven segments: 26 (70 μm along anterior margin), 120, 31, 80, 80, 57, and 37 μm respectively. Two setae on first segment stronger than any others. All setae naked.

Second antenna (Fig. 36f) 374 μm long, 4-segmented and armed as in *M. pertinax*. Fourth segment elongate, 130 μm along outer side, 94 μm along inner side, and 25 μm wide. Two terminal claws slightly unequal, 58 μm and 63 μm.

Labrum, mandible, paragnath, first maxilla, second maxilla, maxilliped, ventral area between maxillipeds and first pair of legs, and legs 1–4 as in *M. pertinax.*

Leg 5 (Fig. 36g) with free segment 129 μm long, its inner margin with blunt proximal process (width of segment here 61 μm) and slight distal swelling (width here 42 μm). Ratio of length to greatest width of free segment 2.11:1. Two terminal elements 57 μm and 91 μm (longer element delicately pectinate along outer margin). Free segment ornamented along outer side with numerous small spines.

Leg 6 represented by two setae on genital area.

Male.—Body (Fig. 37a) slender. Length 1.12 mm (1.01–1.16 mm) and greatest width 0.40 mm (0.38–0.42 mm), based on 5 specimens. Ratio of length to width of prosome 1.75:1. Ratio of length of prosome to that of urosome 1.45:1.

Segment of leg 5 (Fig. 37b) 52 × 109 μm. Genital segment 203 × 192 μm, only slightly longer than wide. Four postgenital segments from anterior to posterior 31 × 75, 31 × 75, 26 × 71, and 44 × 70 μm.

Caudal ramus (Fig. 37c) 44 × 34 μm, shorter than in female, ratio 1.29:1; otherwise resembling that of female.

Rostrum as in female. First antenna segmented and armed as in female, but three long aesthetes added (at positions indicated by dots in Figure 36e), so that formula is same as for male of *M. pertinax.* Second antenna (Fig. 37d) segmented and armed as in female, but showing sexual dimorphism in pectinate nature of setae on first three segments and in having small inner spines on first, second, and fourth segments.

Labrum, mandible, paragnath, first maxilla, second maxilla, maxilliped, ventral area between maxillipeds and first pair of legs, and legs 1–4 as in male of *M. pertinax,* with legs showing similar sexual dimorphism.

Leg 5 (Fig. 37e) with free segment 58 × 14 μm, ratio 4.14:1. Inner terminal element spiniform, 26 μm, and barbed. Outer terminal element 55 μm, setiform, and smooth. Dorsal seta adjacent to free segment about 33 μm and smooth. Free segment ornamented with few minute spines on outer surface.

Leg 6 represented by usual posteroventral flap on genital segment, bearing two slender naked setae 57 μm and 49 μm.

Spermatophore not seen.

Color notes.—Lönning and Vader (in manuscript) describe the color of specimens of *Metaxymolgus turmalis* (their sp. A) as "often off-white or reddish, but they are quite variable in color, as are the sea anemones themselves. The copepods usually have the same color shades as their host and they are therefore easily overlooked. They spend much time in the column of the sea anemones and often emerge on the column when the sea anemone contracts."

Etymology.—The specific name *turmalis,* Latin meaning belonging to the same troop, refers to the apparent close taxonomic relationship of this new species with *M. pertinax* and *M. confinis.*

Remarks.—Although in many respects *Metaxymolgus turmalis* is similar to *Metaxymolgus pertinax,* the new species differs from *M. pertinax* in several ways, principally the relative length of the fourth segment of the second antenna, the length of the caudal ramus, and the form of the free segment of leg 5.

M. *turmalis* may be distinguished from M. *confinis* by the shape of the female genital segment, the length of the female caudal ramus, the two relatively stout setae on the first segment of the first antenna, the length of the fourth segment of the second antenna, and the form of the female leg 5 (see Table 4).

Metaxymolgus sunnivae sp. n.
Figs. 38a–g, 39a–j

Type material.—33 ♀♀, 25 ♂♂, and 10 copepodids from 74 specimens of the actiniarian *Epiactis prolifera* Verrill, low intertidal, Shell Beach, Sonoma County, California, 2–3 December 1979, S. Lönning and W. Vader coll. Holotype ♀, allotype, and 35 paratypes (20 ♀♀, 15 ♂♂) deposited in the National Museum of Natural History, Smithsonian Institution, Washington, D.C.: 20 paratypes (12 ♀♀, 8 ♂♂) deposited in the Zoölogisch Museum, Amsterdam; the remaining paratype (dissected) and the copepodids in the collection of the author.

Other specimens (all from *Epiactis prolifera*).—9 ♀♀, 5 ♂♂, Campbell Cove, Bodega Bay, California, October 1979, S. Lönning and W. Vader coll.; 2 ♀♀, 2 ♂♂, 2 copepodids, Bodega Marine Laboratory, 27 November 1979, S. Lönning and W. Vader coll.; 5 ♀♀, 5 ♂♂, 1 copepodid, Bodega Bay, 11 December 1979, S. Lönning and W. Vader coll.; 1 ♀, 2 ♂♂, intertidal, Cape Arago, Coos Bay, Oregon, 19 June 1980, John Ratliff coll.

From *Tealia crassicornis* (Müller): 1 ♀, intertidal, San Juan Island, Washington, 2 July 1980, W. Vader coll.

From *Tealia lofotensis* (Danielssen): 18 ♀♀, 4 ♂♂, and 12 copepodids, in 4 m, between the jetties, channel of Bodega Harbor, Bodega Bay, California, 9 June 1980, Gregory Ruiz coll.; 66 ♀♀, 53 ♂♂, and 19 copepodids, same locality and date; 3 ♀♀, 1 ♂, and 1 copepodid, off S. jetty, channel of Yaquina Bay, Newport, Oregon, 22 June 1980, John Ratliff coll.

Note.—This species is referred to by Lönning and Vader in manuscript as "*Metaxymolgus* sp. C".

Female.—Body (Fig. 38a) moderately slender. Length 1.57 mm (1.30–1.67 mm) and greatest width 0.62 mm (0.56–0.66 mm), based on 10 specimens. Ratio of length to width of prosome 1.77:1. Ratio of length of prosome to that of urosome 1.74:1.

Segment of leg 5 (Fig. 38b) 91 × 180 μm. Genital segment in dorsal view 234 μm long, broadest in anterior two-thirds (192 μm) and insected at posterior third (125 μm). Genital areas located dorsolaterally near junction of anterior two-thirds of segment. Each area (Fig. 38b, c) with two small naked setae 18 μm and 11 μm. Three postgenital segments from anterior to posterior 78 × 107, 55 × 99, and 81 × 102 μm.

Caudal ramus·81 × 42 μm, ratio 1.93:1, otherwise resembling that of *Metaxymolgus pertinax*.

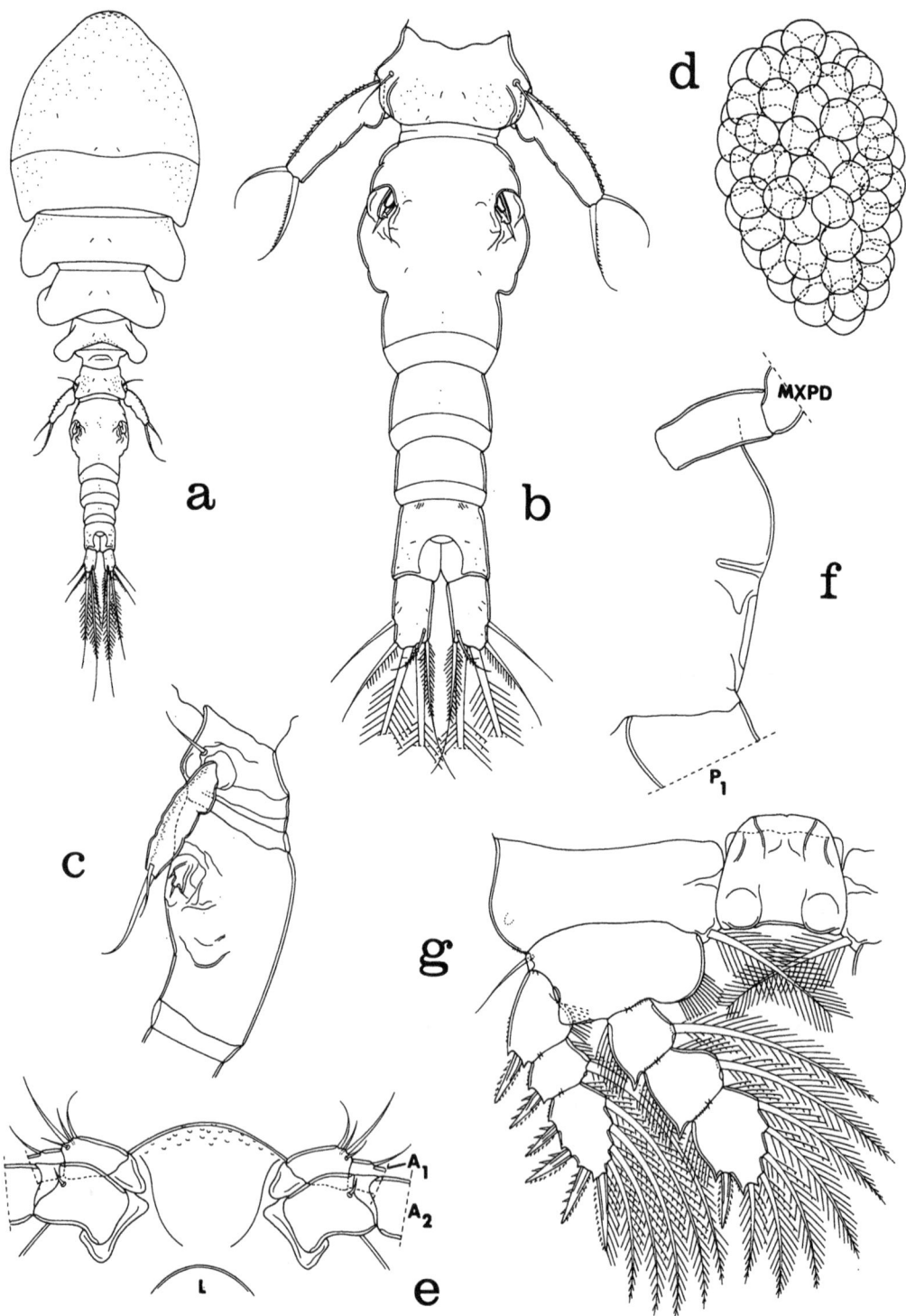

FIG. 38. *Metaxymolgus sunnivae* sp. n., female: a, dorsal (A); b, urosome, dorsal (E); c, segment of leg 5 and genital segment, lateral (E); d, egg sac, ventral (B); e, rostrum, ventral (E); f, area between maxillipeds and first pair of legs, lateral (D); g, leg 1 and intercoxal plate, anterior (D).

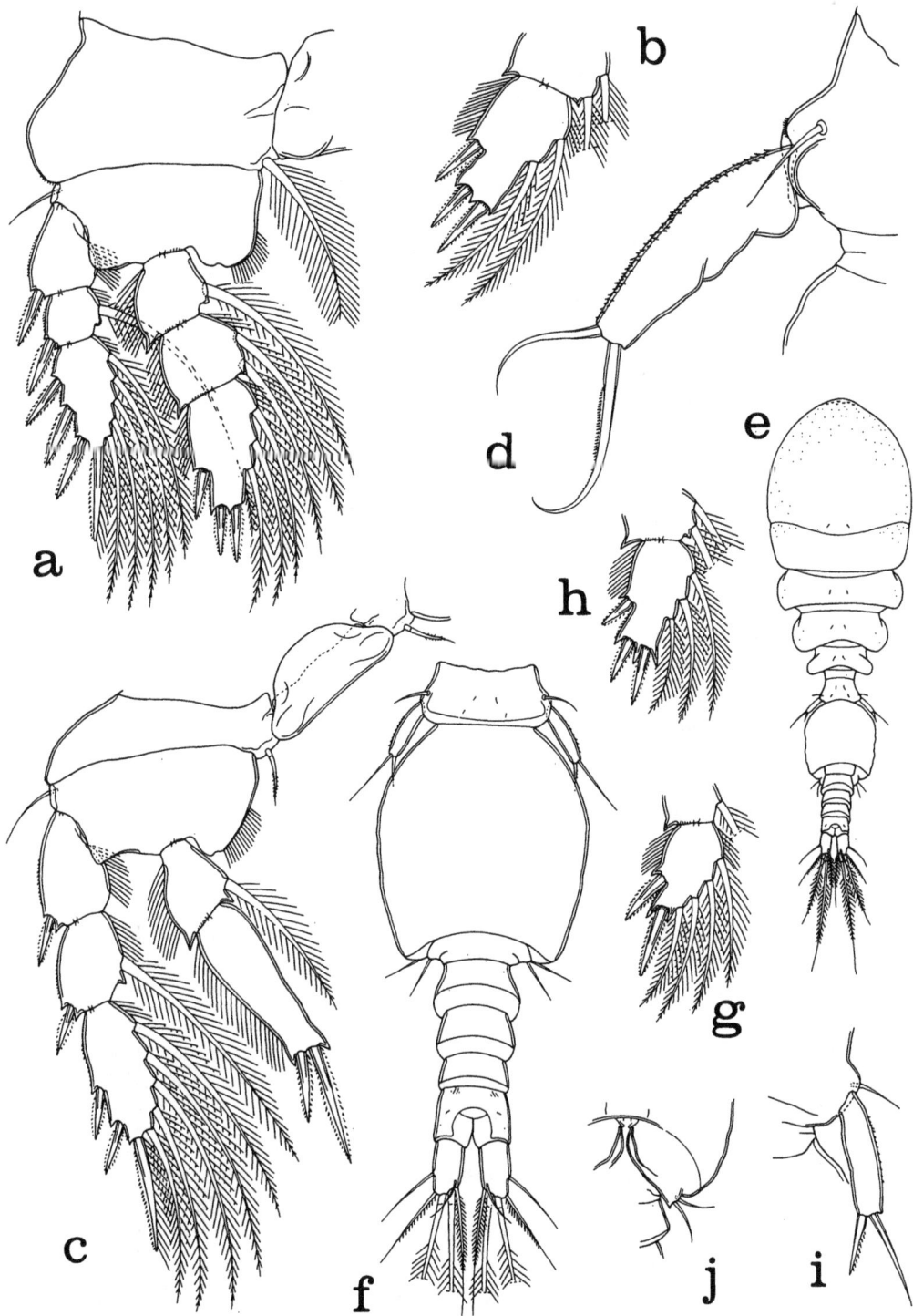

FIG. 39. *Metaxymolgus sunnivae* sp. n. Female: a, leg 2, anterior (D); b, third segment of endopod of leg 3, anterior (D); c, leg 4 and intercoxal plate, anterior (D); d, leg 5, dorsal (F). Male: e, dorsal (A); f, urosome, dorsal (E); g, third segment of endopod of leg 1, anterior (D); h, third segment of endopod of leg 2, anterior (D); i, leg 5, ventral (F); j, leg 6, ventral (E).

Body surface with semilunar markings on front of head, and with numerous refractile points and few hairs (sensilla) as in Figure 35a.

Egg sac (Fig. 38d) oval, 462 × 290 μm, containing numerous eggs about 68–75 μm in diameter.

Rostrum (Fig. 38e) weakly defined posteroventrally.

First antenna similar to that of *M. pertinax*. Length 452 μm. Lengths of seven segments: 31 (73 μm along anterior edge), 114, 33, 73, 72, 55, and 32 μm respectively. Second antenna also similar to that of *M. pertinax*. Length 365 μm. Fourth segment 120 μm along outer side, 75 μm along inner side, and 36 μm wide. Claws 70 μm and 78 μm.

Labrum, mandible, paragnath, first maxilla, second maxilla, and maxilliped resembling those of *M. pertinax*.

Ventral area between maxillipeds and first pair of legs similar to that of *M. pertinax* and projecting slightly (Fig. 38f).

Legs 1–4 (Figs. 38g, 39a–c) segmented and armed as in *M. pertinax*. Leg 4 (Fig. 39c) with inner coxal seta 40 μm and finely barbed, with weak proximal joint as in *M. pertinax*. Exopod 226 μm long. First segment of endopod 52 μm long without spiniform processes (60 μm with these processes) and 45 μm wide, its inner plumose seta 99 μm. Second segment 114 μm long without spiniform processes (121 μm with these processes), greatest width 38 μm, least width 27 μm; two terminal barbed spines 52 μm (outer) and 86 μm (inner).

Leg 5 (Fig. 39d) elongate, free segment 130 μm long, 39 μm wide at level of rounded proximal inner expansion. Two terminal setae 62 μm and smooth, and 96 μm with inner fringe of minute spinules. Dorsal seta approximately 57 μm. Free segment ornamented on outer side with small spines.

Leg 6 represented by two setae on genital segment (Fig. 38b, c).

Male.—Body (Fig. 39e) relatively slender. Length 1.37 mm (1.30–1.43 mm) and greatest width 0.45 mm (0.44–0.50 mm), based on 10 specimens. Ratio of length to width of prosome 1.82:1. Ratio of length of prosome to that of urosome 1.59:1.

Segment of leg 5 (Fig. 39f) 52 × 135 μm. Genital segment in dorsal view 234 × 221 μm, very little longer than wide, with slightly irregular lateral margins. Four postgenital segments from anterior to posterior 52 × 101, 52 × 94, 34 × 81, and 55 × 82 μm.

Caudal ramus as in female but smaller, 68 × 31 μm, ratio 2.19:1.

Rostrum, first antenna, second antenna, labrum, mandible, paragnath, first maxilla, and second maxilla as in male of *M. pertinax*. Maxilliped as in *M. pertinax*, claw 244 μm along its axis.

Ventral area between maxillipeds and first pair of legs as in *M. pertinax*.

Legs 1–4 segmented as in *M. pertinax* with same spine and setal formula as in male of that species. Third segment of endopod of leg 1 (Fig. 39g) with I, I, 4; that of leg 2 (Fig. 39h) with longer terminal spiniform processes than in female.

Leg 5 (Fig. 39i) with slender free segment 61 × 17 μm. Two terminal elements consisting of outer seta 45 μm and inner spiniform element 30 μm. Dorsal seta 30 μm. Free segment with few minute spines along outer edge.

Leg 6 (Fig. 39j) the usual posteroventral flap on genital segment bearing two naked setae about 44 μm long.

Spermatophore not seen.

Color notes.—Lönning and Vader (in manuscript) describe the color of *Metaxymolgus sunnivae* (their sp. C) as tending "to agree in color with their hosts, red sea anemones harboring reddish copepods. The colors of the egg sacs are variable: white, brownish or reddish, more rarely greenish. Different ovigerous females on the same host may have differently hued egg sacs. The copepods on *Epiactis* move all over the host, also over the brooded young. They are extremely difficult to dislodge from their hosts. When the sea anemone contracts the copepods almost always come out on the column."

Etymology.—This species is named for Dr. Sunniva Lönning Vader, University of Tromsö, Tromsö, Norway, who along with her husband, collected the specimens which were generously sent to me for study.

Remarks.—*Metaxymolgus sunnivae* may be distinguished from other species in the genus by a combination of two characters of the female: the shape of the genital segment and the free segment of leg 5 with a rounded inner proximal expansion. Although the new species resembles *Metaxymolgus pertinax* in the general form of many of the appendages, females of the two species may be easily separated by the form of the free segment of leg 5.

Distinctions among the four new species of *Metaxymolgus* are shown in Table 4.

TABLE 4. Comparison of selected features of four new species of *Metaxymolgus*

	M. pertinax	M. confinis	M. turmalis	M. sunnivae
Ratio of greatest length to width of free segment of leg 5	1.11:1	1:1	1.14:1	1.22:1
Ratio of length to width of caudal ramus	1.66:1	2.16:1	1.51:1	1.93:1
Ratio of greatest length to width of fourth segment of second antenna	3.55:1	2.66:1	5.2:1	3.33:1
Ratio of greatest length to width of free segment of leg 5	2.12:1	2.44:1	2.12:1	3.33:1
Proximal inner expansion of leg 5	pointed	pointed	bluntly pointed	broadly rounded

Keys to the species of *Metaxymolgus* associated with Actiniaria

Females

1. Ratio of length to width of caudal ramus more than 1.8:1 2
   Ratio of length to width of caudal ramus less than 1.8:1 3
2. Genital segment as long as wide; free segment of leg 5 with pointed
   proximal inner process . . . . *M. confinis*
   Genital segment longer than wide; free segment of leg 5 with rounded
   proximal inner process . . . . *M. sunnivae*
3. Genital segment longer than wide; free segment of leg 5 with pointed
   proximal inner process 4
   Genital segment as long as wide or wider than long; free segment of
   leg 5 without proximal inner process 5
4. Ratio of caudal ramus 1.66:1; ratio of greatest length to width of
   fourth segment of second antenna 3.55:1; free segment of leg 5 with
   distal rounded process in addition to pointed proximal process
   . . . . *M. pertinax*
   Ratio of caudal ramus 1.51:1; ratio of fourth segment of second antenna
   5.2:1; free segment of leg 5 with only slight swelling in addition to
   pointed proximal process . . . . *M. turmalis*
5. Genital segment as long as wide; ratio of length to width of caudal
   ramus 1.7:1; length of terminal claws of second antenna 60 μm and 35
   μm . . . . *M. cuspis*
   Genital segment wider than long; ratio of length to width of caudal
   ramus 1.24:1; length of terminal claws of second antenna 56 μm and
   42 μm . . . . *M. myorae*

Males

1. One of two setae on second segment of maxilliped highly modified,
   with greatly swollen base armed with spinules 2
   Both setae on second segment of maxilliped unmodified 3
2. Second segment of endopod of leg 4 measuring 101 × 35 μm, the two
   terminal spines being 133 μm and 53 μm; ratio of length of prosome
   to that of urosome 1.9:1 . . . . *M. cuspis*
   Second segment of endopod of leg 4 measuring 112 × 40 μm, the two
   terminal spines being 166 μm and 60 μm; ratio of length of prosome
   to that of urosome 2.2:1 . . . . *M. myorae*
3. Ratio of length to width of caudal ramus 1.59:1 or more; body length
   1.30 mm or more 4
   Ratio of caudal ramus 1.29:1; body length not exceeding 1.16 mm
   . . . . *M. turmalis*
4. Ratio of caudal ramus 2.19:1; leg 5 with ratio of length to width less
   than 4:1 5
   Ratio of caudal ramus 1.59:1; leg 5 with ratio of length to width 4.94:1
   . . . . *M. pertinax*
5. Genital segment somewhat globose with smoothly rounded lateral
   margins; second antenna moderately robust, with ratio of greatest

length to width of fourth segment being 3.13:1 .... *M. confinis*
Genital segment slightly subquadrate with lateral margins a little flat-
tened and irregular; second antenna elongate, ratio of greatest length
to width of fourth segment being 3.59:1 .... *M. sunnivae*

Genus *Notoxynus* Humes, 1975
Notoxynus crinitus sp. n.
Figs. 40a–i, 41a–j, 42a–f

Type material.—4 ♀♀, 7 ♂♂ from one actiniarian, *Heteractis crispa* (Ehrenberg), in 3 m, north of Isle Maître, near Nouméa, New Caledonia, 22°19′30″S,
166°24′35″E, 13 July 1971. (This is the same host individual from which
Doridicola cylichnophorus, D. paterellis, D. scyphulanus, and *D. dunnae* were
recovered). Holotype ♀, allotype, and 5 paratypes (1 ♀, 4 ♂♂) deposited in
the National Museum of Natural History, Smithsonian Institution, Washington, D.C.; the remaining paratypes (dissected) in the collection of the
author.

Other specimens.—1 ♀, 1 ♂ from one *Heteractis crispa,* in 0.5 m, western
side of Isle To N'du, southwest of Pte. Laguerre, 10 km southwest of Paita,
New Caledonia, 22°10′42″S, 166°16′30″E, 29 June 1971.

Female.—Body (Fig. 40a) moderately slender. Length 2.10 mm (1.92–2.24
mm) and greatest width 0.72 mm (0.67–0.78 mm), based on four specimens.
Segment of leg 1 separated from cephalosome by dorsal transverse furrow.
Epimera of legs 1–4 rounded posteriorly. Ratio of length to width of prosome
1.49:1. Ratio of length of prosome to that of urosome 1.03:1.

Segment of leg 5 (Fig. 40b) 122 × 250 μm. Genital segment elongate, 221
μm long, 216 μm wide in anterior half and slightly narrower in posterior
half. Genital areas located dorsolaterally in anterior half of segment. Each
area (Fig. 40c) with two small naked setae approximately 8 μm and 13 μm
and a spiniform process. Three postgenital segments from anterior to posterior 114 × 138, 96 × 133, and 135 × 120 μm. Posteroventral border of anal
segment with conspicuous spiniform process and several small spines
(Fig. 40d).

Caudal ramus (Fig. 40d) elongate, 273 × 49 μm, ratio of length to width
5.57:1. Outermost terminal seta 99 μm, innermost terminal seta 156 μm.
Outer lateral seta 112 μm, dorsal seta 65 μm. Outer median terminal seta
143 μm. Inner median terminal seta broken in all specimens but longer
than outer median terminal seta. All setae proximally with long hairs
(length up to 24 μm) except smooth dorsal seta and inner median
terminal seta.

Body surface with few hairs (sensilla) as in Figure 40a.

Egg sac (Fig. 40e) elongate oval, 528 × 305 μm, containing 17 eggs about
130–143 μm in diameter.

Rostrum (Fig. 40f) rounded posteroventrally.

First antenna (Fig. 40g) 448 μm long. Lengths of seven segments: 36 (73
μm along anterior margin), 112, 43, 78, 65, 43, and 34 μm respectively.
Formula for armature: 4, 13, 6, 3, 4 + 1 aesthete, 2 + 1 aesthete, and 7

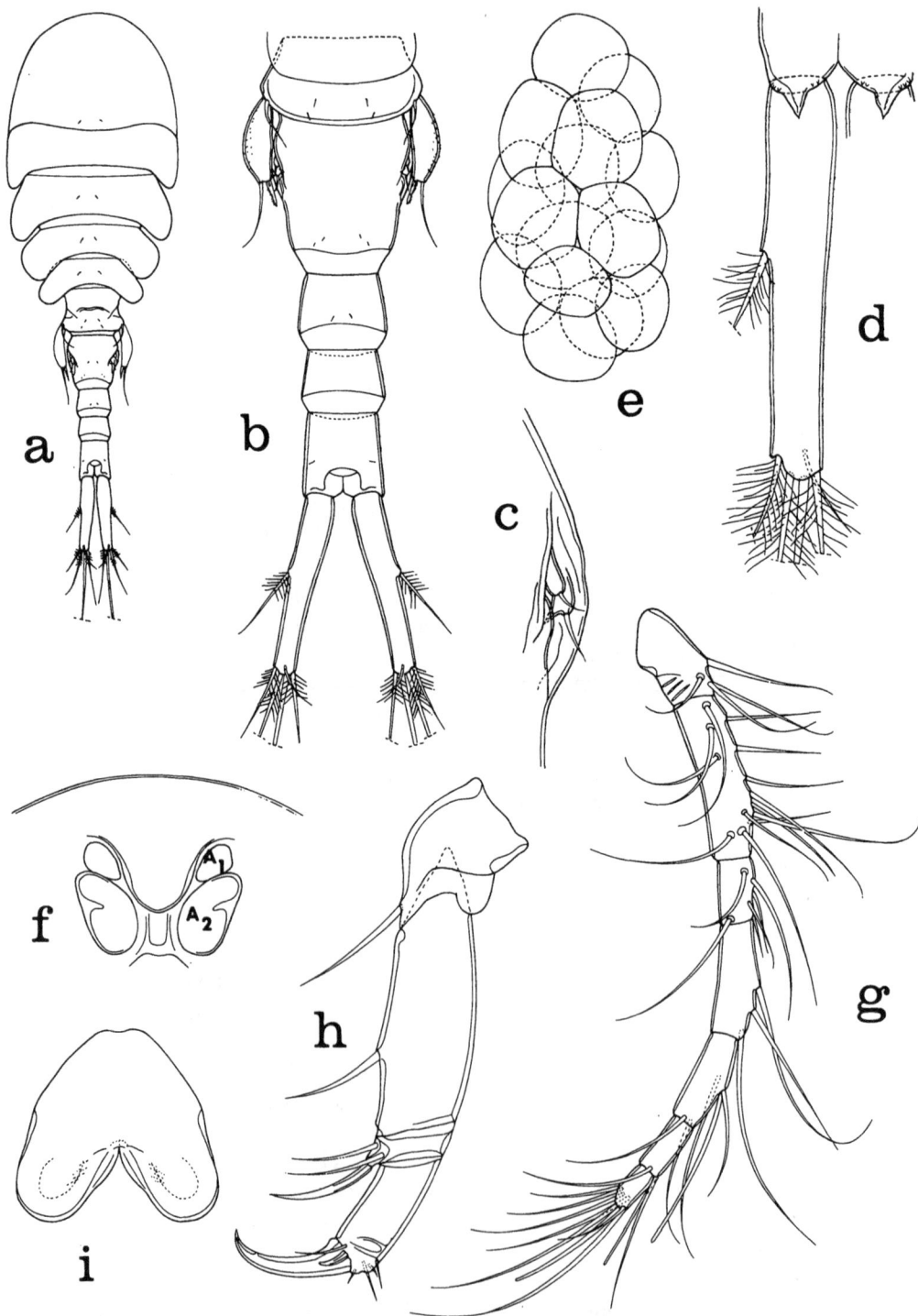

FIG. 40. *Notoxynus crinitus* sp. n., female: a, dorsal (I); b, urosome, dorsal (B): c, genital area, dorsal (C); d, caudal ramus, ventral (D); e, egg sac, ventral (B); f, rostrum, ventral (B); g, first antenna, dorsal (D); h, second antenna, posterior (F); i, labrum, with paragnaths indicated by broken lines, ventral (D).

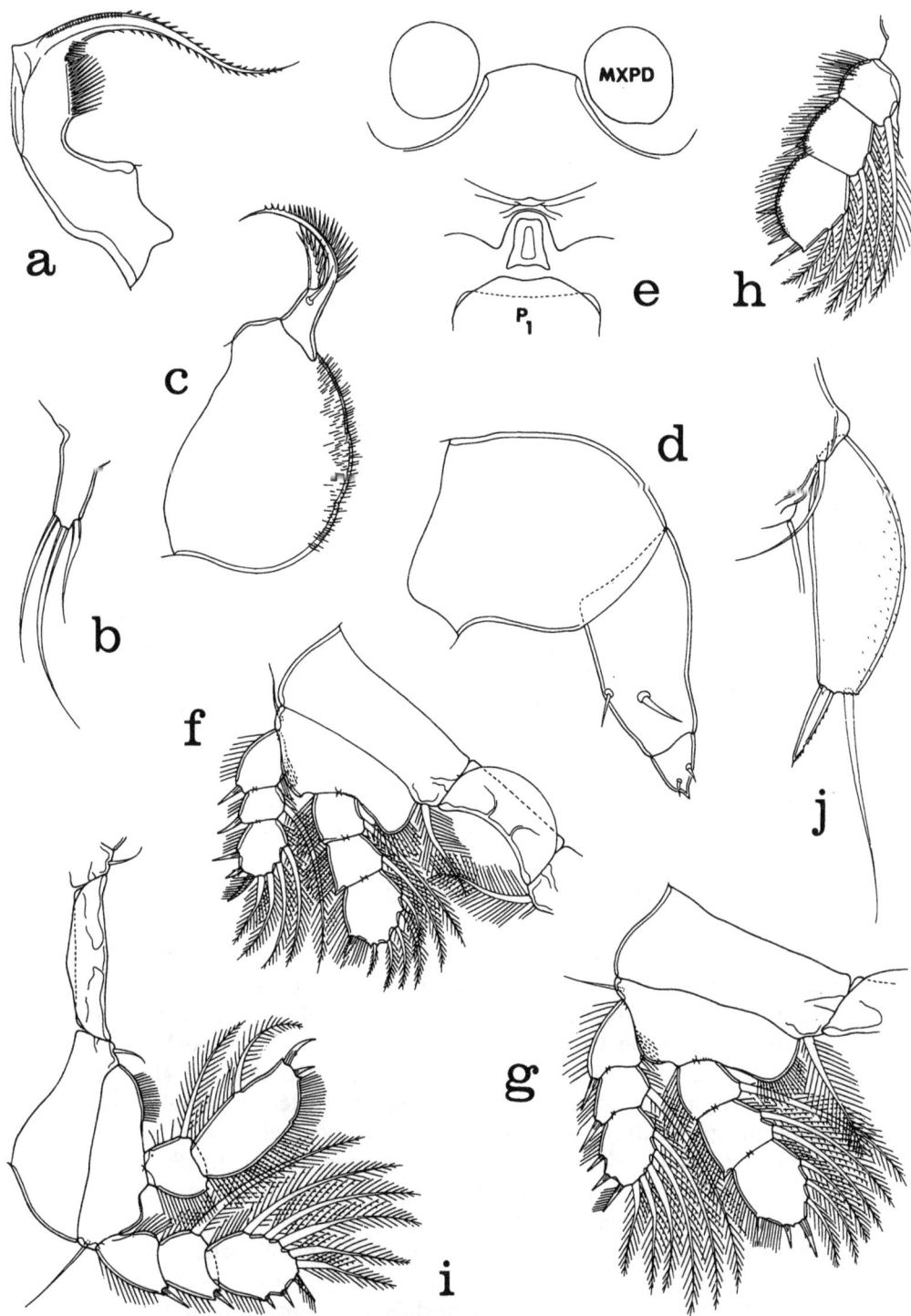

FIG. 41. *Notoxynus crinitus* sp. n., female: a, mandible, posterior (C); b, first maxilla, posterior (C); c, second maxilla, posterior (F); d, maxilliped, postero-inner (F); e, area between maxillipeds and first pair of legs, ventral (E); f, leg 1 and intercoxal plate, anterior (E); g, leg 2, anterior (E); h, endopod of leg 3, posterior (E); i, leg 4 and intercoxal plate, anterior (E); j, leg 5, dorsal (F).

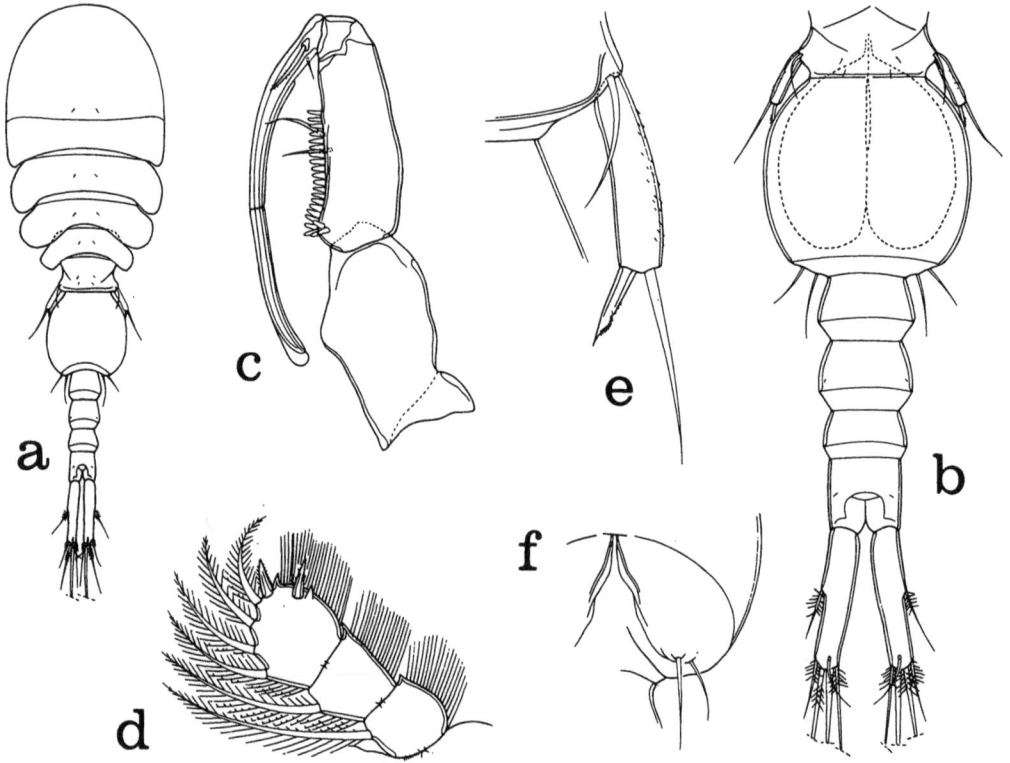

FIG. 42. *Notoxynus crinitus* sp. n., male: a, dorsal (I); b, urosome, dorsal (B); c, maxilliped, inner (E); d, endopod of leg 1, anterior (D); e, leg 5, dorsal (F); f, leg 6, ventral (E).

+ 1 aesthete. All setae naked. First segment with slender oblique sclerotized bars.

Second antenna (Fig. 40h) 4-segmented and 257 μm long including claw, with formula 1, 1, 3, and I + 6 (one of these elements stout and clawlike, 15 μm long). Fourth segment 66 μm along outer side, 42 μm along inner side, and 26 μm wide. Claw 61 μm along its axis. All setae naked.

Labrum (Fig. 40i) with two posteroventral lobes. Mandible (Fig. 41a) with concave margin beyond indentation bearing row of slender spinules; with convex margin lacking spinules and leading to a pointed apex followed by a minutely striated fringe. Lash long and barbed. Paragnath (Fig. 40i) a small lobe with few hairs. First maxilla (Fig. 41b) with three setae. Second maxilla (Fig. 41c) 2-segmented, first segment swollen with outer hairs. Second segment slender, bearing posterior surficial smooth seta, and inner barbed (dorsal) seta; lash long with graduated spinules. Maxilliped (Fig. 41d) 3-segmented. First segment unarmed. Second segment with two short smooth setae. Small third segment with two small setae and terminating in sharp point.

Ventral area between maxillipeds and first pair of legs (Fig. 38e) not protuberant.

Legs 1–4 (Fig. 41f–i) with 3-segmented rami except for 2-segmented endopod of leg 4. Formula for arrangement of spines and setae as in *Notoxynus mundus* Humes, 1975, except for third exopod segment of legs 1–3 having 3 spines instead of 4 as in that New Caledonian species. Outer margins of exopod segments in all four legs with delicate hairs. Proximal few outer spines on exopods in these legs smooth rather than barbed. Third endopod segment in legs 1–3 with outer hairs continuing distally beyond spine. Leg 4 with exopod 211 μm long. First endopod segment 55 $\times$ 62 μm. Second segment 140 $\times$ 73 μm, its two terminal spines 19 μm and 40 μm. Inner coxal seta of leg 4 33 μm and smooth.

Leg 5 (Fig. 41j) with free segment 125 $\times$ 49 μm. Terminally with unilaterally barbed spine 39 μm (44 μm including terminal filament) and smooth seta 112 μm. Free segment ornamented with minute spinules along outer surface. Dorsal seta approximately 80 μm and naked.

Leg 6 represented by two setae and spiniform process on genital area (Fig. 40c).

Color in living specimens in transmitted light very pale brown, eye red, egg sacs gray.

Male.—Body (Fig. 42a) slender. Length 1.87 mm (1.73–2.02 mm) and greatest width 0.59 mm (0.54–0.64 mm), based on seven specimens. Ratio of length to width of prosome 1.49:1. Ratio of length of prosome to that of urosome 1:0.88, urosome being a little longer than prosome.

Segment of leg 5 (Fig. 42b) 65 $\times$ 198 μm. Genital segment 250 $\times$ 290 μm, a little wider than long. Four postgenital segments from anterior to posterior 86 $\times$ 135, 91 $\times$ 127, 65 $\times$ 114, and 104 $\times$ 104 μm.

Caudal ramus resembling that of female but shorter, 185 $\times$ 44 μm, ratio 4.20:1.

Rostrum, first antenna, second antenna, labrum, mandible, paragnath, first maxilla, and second maxilla as in female. Maxilliped (Fig. 42c) 4-segmented (assuming that proximal part of claw represents fourth segment). First segment unornamented. Second segment with two slender naked setae and row of stout spines on inner surface. Third segment small. Claw 330 μm, with subdivision midway and two unequal proximal setae, longer of these minutely barbed.

Ventral area between maxillipeds and first pair of legs as in female.

Legs 1–4 segmented and armed as in female, except for third endopod segment of leg 1 where formula is I, I, 4 (Fig. 42d) instead of I, 5 as in female.

Leg 5 (Fig. 42e) with slender free segment, 88 $\times$ 21 μm.

Leg 6 (Fig. 42f) a posteroventral flap on genital segment bearing two slender naked setae 83 μm and 65 μm.

Extruded spermatophore not seen.

Color in living specimens as in female.

Etymology.—The specific name *crinitus*, Latin meaning with long hairs, alludes to the unusually long hairs on the rami of legs 1–4.

Remarks.—The genus *Notoxynus* previously contained only a single species, *Notoxynus mundus* Humes, 1975, from the alcyonacean *Xenia membranacea* Schenk in New Caledonia. *Notoxynus crinitus* may be easily distinguished from *N. mundus* by the small free segment of leg 5 in the latter. Other differences are to be seen in the length of the caudal ramus, the nature of the armature on the lash of the second maxilla, and the shape of the second endopod segment of leg 4. Distinctive features of the two species are summarized in Table 5.

TABLE 5. Comparison of females of the two species of *Notoxynus*

	Notoxynus crinitus	*Notoxynus mundus*
Caudal ramus	ratio 5.57:1, with long proximal hairs on certain setae	ratio 4:1, all setae naked
Posteroventral border of anal segment	with a conspicuous spiniform process	smooth
Outer margin of exopod segments of legs 1–4	with long hairs	smooth
Second segment of endopod of leg 4	140 × 73 μm, 1.92:1, stout	117 × 47 μm, 2.49:1, slender
Free segment of leg 5	large, 125 × 49 μm	small, 33 × 22 μm

Genus *Paramolgus* Humes and Stock, 1973

Paramolgus antillianus Stock, 1975a

Host: *Ricordea florida* Duchassaing and Michelotti.
Site: In washings.
Locality: Cayo Enrique, off La Parguera, Puerto Rico (Stock, 1975a).
Notes: Length of ♀ 1.10 mm, ♂ 0.77 mm.

Paramolgus politus (Humes and Ho, 1967)
= *Lichomolgus politus* Humes and Ho, 1967

Host: *Rhodactis rhodostoma* (Ehrenberg).
Site: In washings.
Locality: Region of Nosy Bé, northwestern Madagascar (Humes and Ho, 1967).
Notes: Length of ♀ 1.78 mm, ♂ 1.33 mm.

Paramolgus simulans (Humes and Ho, 1967)
= *Lichomolgus simulans* Humes and Ho, 1967

Host: *Rhodactis rhodostoma* (Ehrenberg).
Site: In washings.
Locality: Region of Nosy Bé, northwestern Madagascar (Humes and Ho, 1967).
Notes: Length of ♀ 1.40 mm, ♂ 0.96 mm.

Verutipes gen. n.

Diagnosis.—Lichomolgidae. Body elongate. Cephalosome distinctly broader than rest of body. Urosome 5-segmented in female, 6-segmented in male. Caudal ramus with six setae. Rostrum not developed. First antenna 7-segmented. Second antenna 4-segmented with two terminal claws.

Labrum with two posteroventral lobes. Mandible having on convex edge a scalelike area with row of spinules followed by smooth area and then a serrated fringe. Concave margin beyond indentation bearing row of long spinules. Lash long. Paragnath a small lobe. First maxilla with three setae. Second maxilla 2-segmented. Maxilliped 3-segmented in female, 4-segmented in male if claw is considered as a segment.

Legs 1 and 2 with 3-segmented rami. Legs 3 and 4 with 3-segmented exopods and 2-segmented endopods, endopods of both legs with formula 0-0; I.

Leg 5 placed ventrally, with small free segment indistinctly set off from body segment, bearing two terminal setae.

Other features as in species below.

Associated with actiniarians.

Gender masculine.

Type-species.—*Verutipes laticeps* sp. n.

Etymology.—The generic name is a combination of Latin *verutus*, armed with a dart or javelin, and *pes*, foot, alluding to the single spine on the tip of the endopod in legs 3 and 4.

Verutipes laticeps sp. n.
Figs. 43a–f, 44a–k, 45a–h

Type material.—10 ♀♀, 4 ♂♂ from 12 specimens of the actiniarian *Entacmaea quadricolor* (Rueppell and Leuckart), in 1 m, Isle Mando, near Nouméa, New Caledonia, 22°18'59"S, 166°09'30"E, 5 July 1971. Holotype ♀, allotype, and 8 paratypes (7 ♀♀, 1 ♂) deposited in the National Museum of Natural History, Smithsonian Institution, Washington, D.C.; the remaining paratypes (dissected) in the collection of the author.

Other specimens (all from *Entacmaea quadricolor*).—2 ♀♀ from 3 hosts, in 1 m, western side of Isle Maître, near Nouméa, 22°20'05"S, 166°24'05"E, 11 June 1971; 1 ♂ from 15 hosts, in tide pool, eastern side of Isle Maître, 22°20'35"S, 166°25'10"E, 16 July 1971; 1 ♀, 1 ♂ from 3 hosts, in 1 m, western end of Isle Mando, 22°18'59"S, 166°09'30"E, 3 July 1971.

Female.—Body (Fig. 43a, b) elongate, with cephalosome wider than metasome. Sides of cephalosome with three lobes. Posterior part of prosome with pair of posterolateral lobes. Urosome relatively slender. Length of body 0.95 mm (0.73–1.03 mm) and greatest cephalosomal width 0.44 mm (0.40–0.48 mm), based on 10 specimens. Width of prosome at level of segment of fourth leg 308 μm. Dorsoventral thickness of body at level of mouth parts 340 μm, at level of second leg 264 μm. Prosome with ratio of length to greatest width (at level of cephalosome) 1.56:1. Ratio of length of prosome to that of urosome 2.5:1.

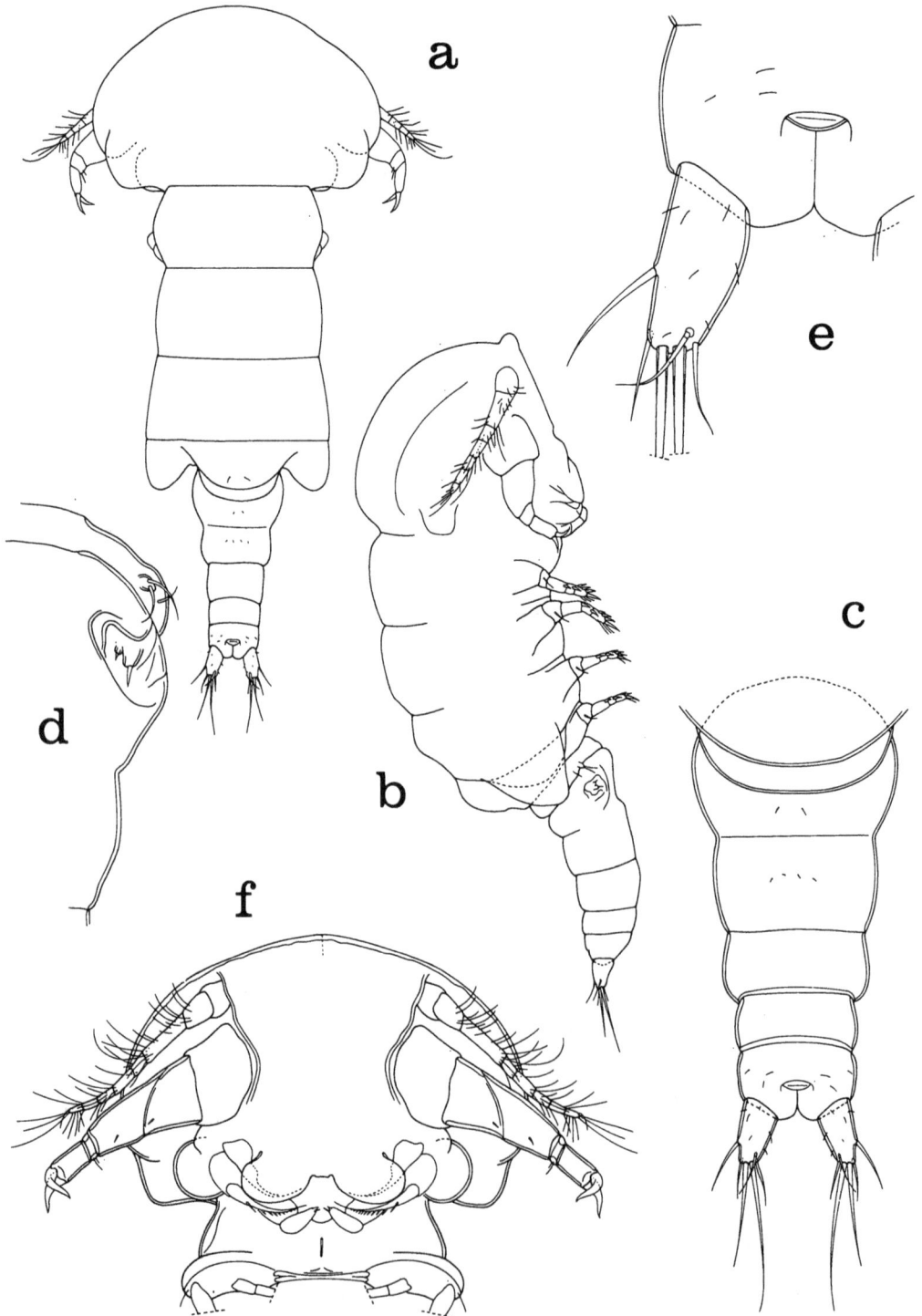

FIG. 43. *Verutipes laticeps* gen. n., sp. n., female: a, dorsal (B); b, lateral (B); c, urosome, dorsal (D); d, leg 5 and genital area, ventral (F); e, caudal ramus, dorsal (G); f, cephalosome, ventral (E).

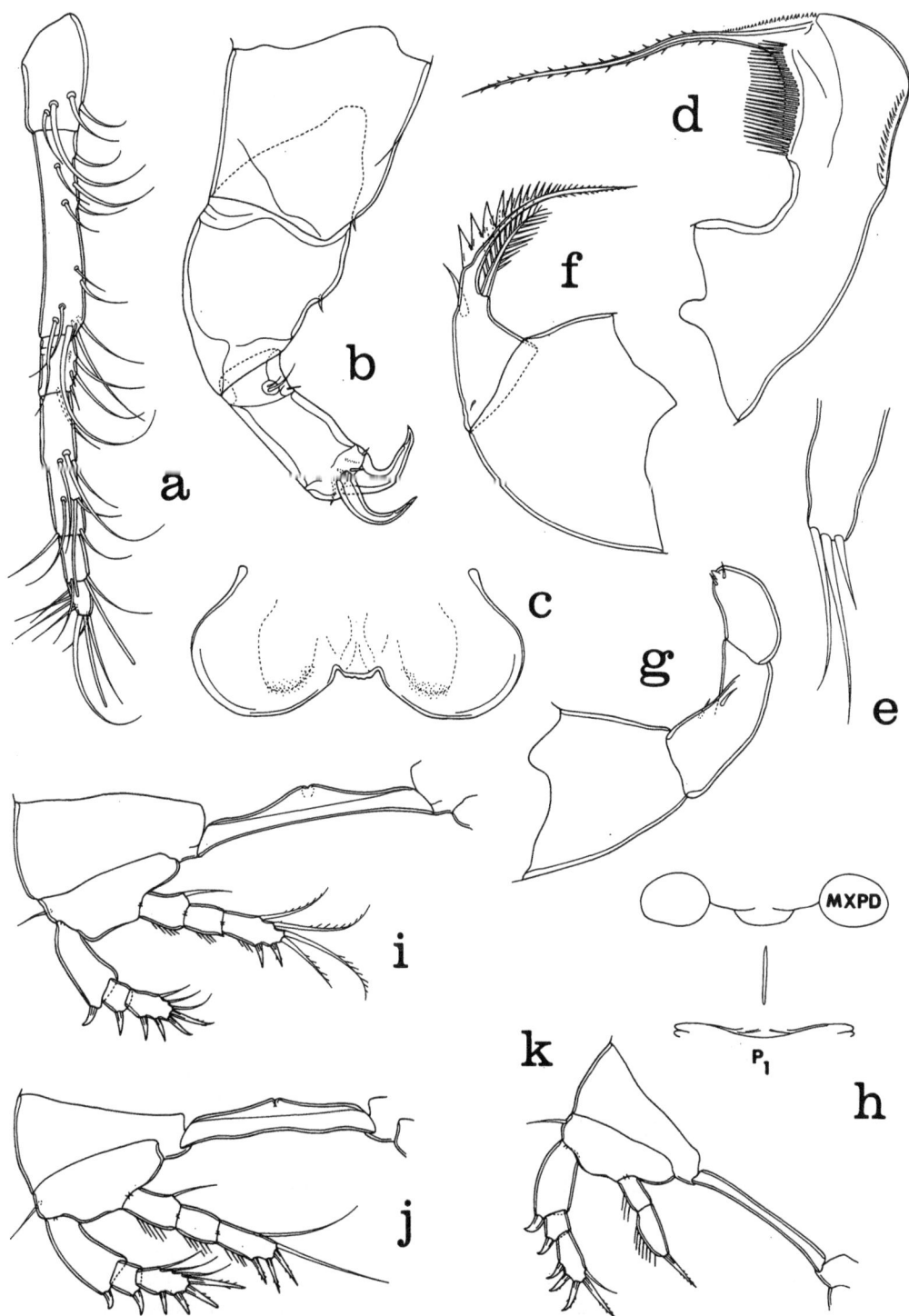

FIG. 44. *Verutipes laticeps* gen. n., sp. n., female: a, first antenna, ventral (C); b, second antenna, postero-inner (C); c, labrum, with paragnaths indicated by broken lines, ventral (C); d, mandible, posterior (H); e, first maxilla, posterior (H); f, second maxilla, anterior (G); g, maxilliped, postero-inner (G); h, area between maxillipeds and first pair of legs, ventral (F); i, leg 1 and intercoxal plate, anterior (F); j, leg 2 and intercoxal plate, anterior (F); k, leg 3 and intercoxal plate, posterior (F).

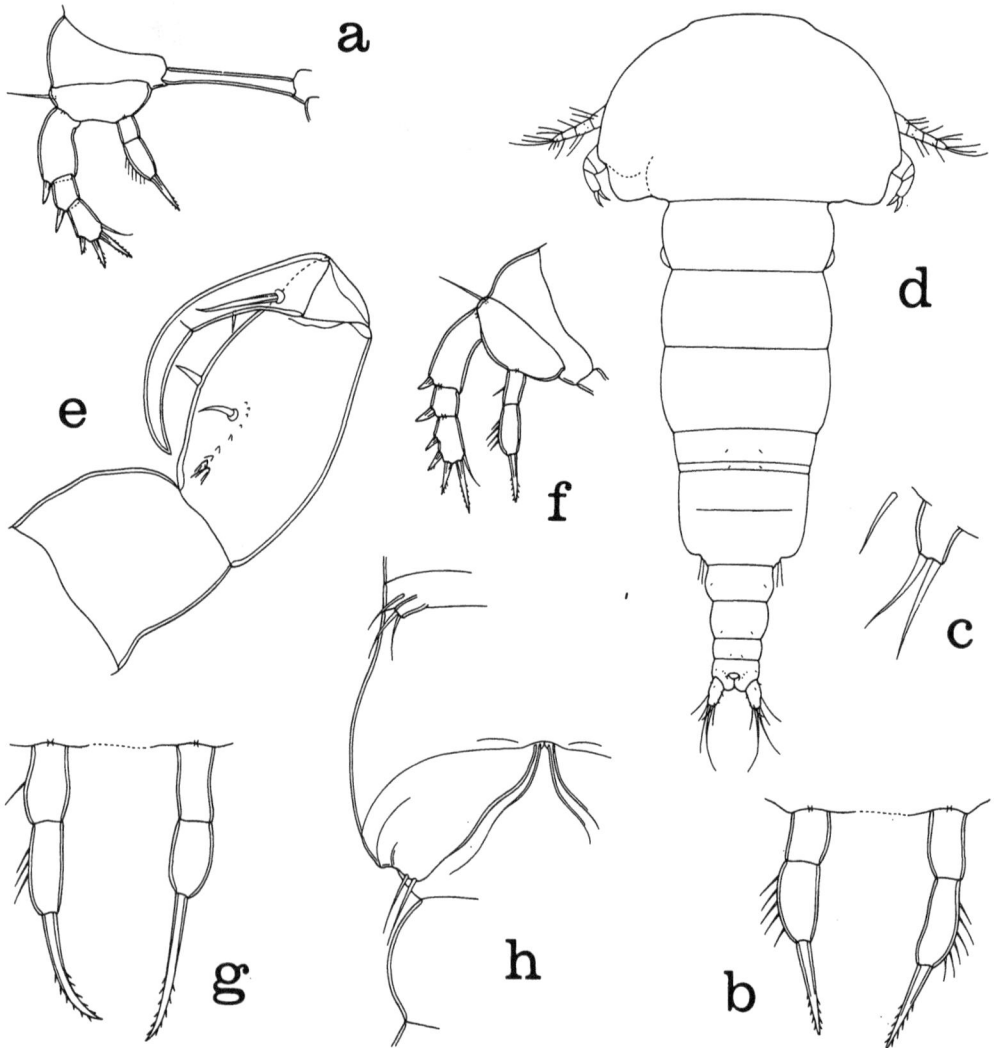

FIG. 45. *Verutipes lacticeps* gen. n., sp. n. Female: a, leg 4 and intercoxal plate, anterior (F); b, endopods of fourth pair of legs in one female, anterior (G); c, leg 5, lateral (G). Male: d, dorsal (B); e, maxilliped, inner (G); f, leg 4, anterior (F); g, endopods of fourth pair of legs in one male, anterior (G); h, leg 5 and leg 6, ventral (F).

Segment of leg 5 (Fig. 43c) short dorsally, approximately 18 × 156 μm, partly covered dorsally by tergum of segment of leg 4. Genital segment with slight hourglass form, with transverse line dorsally at level of constriction but this line not continued ventrally. Length of genital segment 158 μm (anterior part 96 μm, posterior part 62 μm). Width in anterior part 159 μm and in posterior part 120 μm. Genital areas located ventrolaterally (Fig. 43d). Each area with two minute setae 5 μm and 9 μm. Three postgenital segments from anterior to posterior 65 × 112, 39 × 94, and 50 × 88 μm. Posteroventral margin of anal segment smooth.

Caudal ramus (Fig. 43e) 46×22 μm, ratio 2.01:1. Outer lateral seta 35 μm. Dorsal seta 28 μm. Outermost terminal seta 24 μm and innermost terminal seta 28 μm. Two median terminal setae 61 μm (outer) and 84 μm (inner). All setae naked. Dorsal surface of ramus with few hairs.

Body surface with very few hairs (sensilla) as in Figure 43a.

Egg sac not seen.

Rostrum (Fig. 43f) without well–defined posteroventral margin.

First antenna (Fig. 44a) 203 μm long. Lengths of seven segments: 24 (39 μm along anterior margin), 71, 21, 24, 23, 14, and 11 μm respectively. Formula for armature: 4, 13, 6, 3, $4 + 1$ aesthete, $2 + 1$ aesthete, and $7 + 1$ aesthete. All setae smooth.

Second antenna (Fig. 44b) 164 μm long without claws. Formula: 1, 1, 3, and 2 claws plus 5 setules. Fourth segment 55 μm along outer side, 33 μm along inner side, and 22 μm wide. One claw 34 μm and slightly sinuous, other claw 27 μm and less sinuous.

Labrum (Fig. 44c) with two broad rounded diverging lobes. Mandible (Fig. 44d) with convex side having scalelike area with row of spinules followed by smooth hyaline region and then by finely serrate fringe. Concave side beyond indentation with row of long spinules. Lash long and barbed. Paragnath (Fig. 44c) a small hairy lobe. First maxilla (Fig. 44e) with three setae. Second maxilla (Fig. 44f) 2-segmented with usual lichomolgid armature. Proximal teeth on lash with sharply pointed tips. Maxilliped (Fig. 44g) 3-segmented with armature as in other lichomolgids.

Ventral area between maxillipeds and first pair of legs not protuberant (Fig. 44h).

Legs 1–4 (Figs. 44i–k, 45a) with 3-segmented rami except for 2-segmented endopods in legs 3 and 4. Formula for armature as follows (Roman numerals indicating spines, Arabic numerals representing setae):

P_1	coxa	0-0	basis	1-0	exp	I-0;	I-0;	III, I, 3
					enp	0-1;	0-0;	II, 2, 2
P_2	coxa	0-0	basis	1-0	exp	I-0;	I-0;	III, I, 3
					enp	0-1;	0-0;	II, I, 1, 1
P_3	coxa	0-0	basis	1-0	exp	I-0;	I-0;	III, I, 1
					enp	0-0;	I	
P_4	coxa	0-0	basis	1-0	exp	I-0;	I-0;	III, I, 1
					enp	0-0;	I	

In all four legs coxa lacking inner seta and inner margin of basis smooth. Spine on exopods with tips recurved posteriorly. Legs 1 and 2 with second segment of endopod unarmed. Legs 3 and 4 with 2-segmented endopods, first segment unarmed, second segment with single terminal barbed spine. Endopod of leg 4 showing variation in form (Fig. 45b) and spine about 19 μm long.

Leg 5 (Fig. 45c) small and placed ventrally (Fig. 43d). Free segment, incompletely set off from body segment, approximately 8×6 μm, with two terminal setae 22 μm. Dorsal seta, in this case moved ventrally, about 12 μm.

Leg 6 probably represented by two small setae on genital area (Fig. 43d).

Color in living specimens in transmitted light opaque gray, eye red.

Male.—Body (Fig. 45d) with broad cephalosome as in female but lacking posterolateral lobes on posterior part of prosome seen in that sex. Length of body 0.86 mm (0.61–0.95 mm) and greatest width 0.37 mm (0.30–0.41 mm), based on 7 specimens. Greatest width of cephalosome 429 μm. Width at level of segment of third legs 239 μm. Prosome with ratio of length to greatest width (at level of cephalosome) 1.38:1. Ratio of length of prosome to that of urosome 1.8:1.

Segment of leg 5 (Fig. 45d) short, 15 × 185 μm. Genital segment 117 × 187 μm, wider than long. Four postgenital segments from anterior to posterior 52 × 100, 44 × 83, 29 × 70, and 34 × 70 μm.

Caudal ramus as in female, but smaller, 40 × 21 μm, ratio 1.90:1.

Body surface with few hairs (sensilla) as in Figure 45d.

Rostrum, first antenna, second antenna, labrum, mandible, paragnath, first maxilla, and second maxilla as in female. Maxilliped (Fig. 45e) 4-segmented, assuming proximal part of claw to represent fourth segment. First segment unarmed. Second segment with two setae and row of graduated spines. Small third segment unarmed. Claw relatively short, 57 μm along its axis, weakly subdivided at midlength, and bearing proximally two unequal smooth setae.

Ventral area between maxillipeds and first pair of legs as in female.

Legs 1–4 segmented and armed as in female. One male with third exopod segment of both fourth legs having formula III, I, 1 (Fig. 45f) instead of usual II, I, 1. Endopod of leg 4 as in female, but one male having endopods with relatively longer spines as in Figure 45g.

Leg 5 (Fig. 45h) as in female.

Leg 6 (Fig. 45h) a posteroventral flap on genital segment bearing two setae 31 μm and 35 μm.

Spermatophore not seen.

Color as in female.

Etymology.—The specific name *laticeps*, a combination of *latus*, broad, and *ceps*, head, refers to the unusually broad cephalosome.

Remarks.—*Verutipes laticeps* differs from all other Lichomolgidae in having the endopods of both leg 3 and leg 4 with the formula 0-0; I. The unusually broad cephalosome is also distinctive in this species.

Family Sabelliphilidae Gurney, 1927
Genus *Paranthessius* Claus, 1889

Paranthessius anemoniae Claus, 1889

Host: *Anemonia* sp.
Site: Not given.
Locality: Trieste, Italy (Claus, 1889).

Host: *Anemonia sulcata* (Pennant).

Site: On the column or, if the anemone is detached, on the pedal disk (Bocquet and Stock, 1959).

Localities: Trieste, Italy (Graeffe, 1900); Strangford Lough, Northern Ireland (Gotto and Briggs, 1972); vicinity of Roscoff, France (Bocquet and Stock, 1959).

Notes: Length of ♀ 1.5–1.6 mm, ♂ 0.95–1 mm (Bocquet and Stock, 1959).

Citations: Zulueta (1911), Bouligand (1966).

Experimental host: *Actinia equina* Linnaeus. See Briggs (1976).

Siphonostomatoida Kabata, 1979
Family Asterocheridae Giesbrecht, 1899

Genus Asterocheres Boeck, 1859
Asterocheres scutatus Stock, 1966

Host: *Rhodactis rhodostoma* (Ehrenberg).

Site: Unknown.

Locality: Eilat, Gulf of Aqaba (Stock, 1966).

Notes: Length of ♀ 0.55 mm, ♂ unknown.

Genus *Asteropontius* Thompson and A. Scott, 1903

Asteropontius longipalpus Stock, 1975a

Host: *Ricordea florida* Duchassaing and Michelotti.

Site: Not given.

Localities: Cayo Enrique, off La Parguera, Puerto Rico (Stock, 1975a); Jamaica (present paper).

Notes: Length of ♀ 1.16 mm (Stock, 1975a), ♂ 0.74 mm (present paper).

In Stock's original description of *Asteropontius longipalpus* only the female was included, since males had not at that time been discovered. Males have now been found in a collection made by the author from *Ricordea florida* in Jamaica in 1959. A brief description of these males follows.

Material examined.—10 ♀♀, 4 ♂♂, 1 copepodid from several corallimorpharians, *Ricordea florida* Duchassaing and Michelotti, in 1 m, surf zone of outer reef, South East Cay, Jamaica, 6 September 1959.

Male.—Body as Stock (1975a) has figured for *Asteropontius parvipalpus*. Length 0.74 mm (0.73–0.75 mm) and greatest width 0.31 mm (0.30–0.31 mm), based on 4 specimens.

Segment of leg 5 (Fig. 46a) 36 × 109 μm. Genital segment 104 × 120 μm, a little wider than long. Three postgenital segments from anterior to posterior 40 × 62, 39 × 57, and 29 × 55 μm. Caudal ramus 26 × 24 μm, only a very little longer than wide.

First antenna (Fig. 46b) with 17 segments (segment 10 short and obscure), their lengths: 29 (39 μm along anterior margin), 15.5, 12.5, 10, 10, 9, 11, 11, 6.5, 3, 10, 42, 22, 19, 41, 34, and 24 μm respectively. Formula for armature:

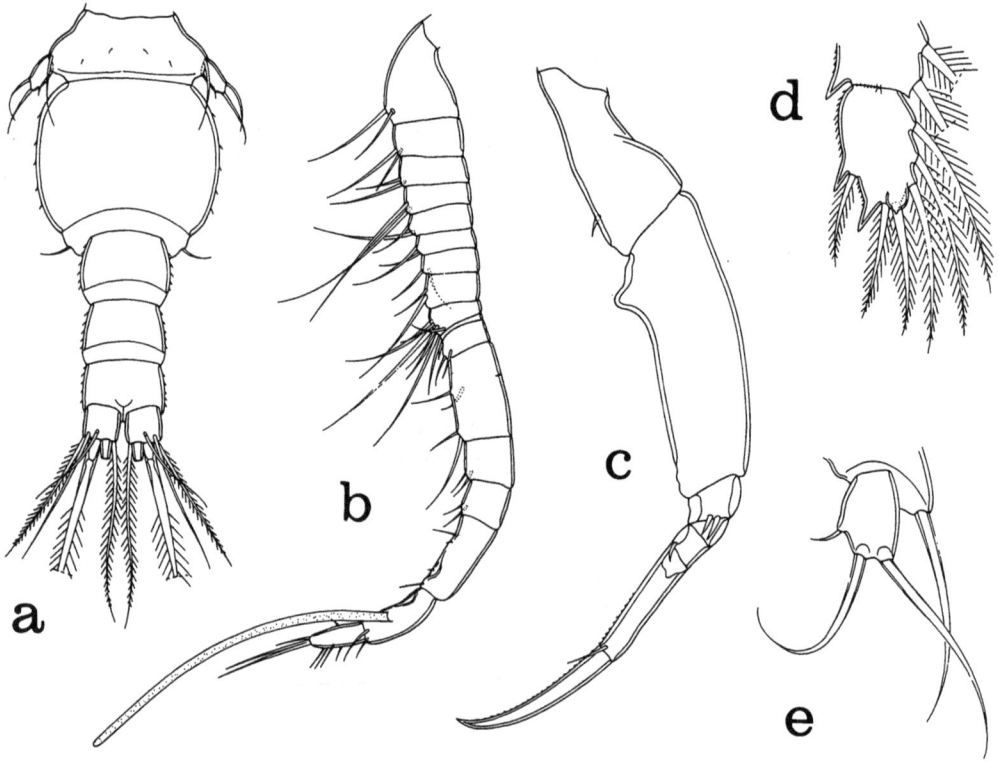

FIG. 46. *Asteropontius longipalpus* Stock, 1975, male: a, urosome (D); b, first antenna, ventral (F); c, maxilliped, postero-inner (C); d, third segment of endopod of leg 1, anterior (C); e, leg 5, lateroventral (G).

2, 2, 2, 2, 2, 2, 2, 2, 7, 2, 2, 4 (2 + 2), 2, 2, 4 (2 + 2), 2, and 8. Aesthete on penultimate segment.

Second antenna, mandible, siphon, first maxilla, and second maxilla as in female (see Stock's description). Maxilliped (Fig. 46c) resembling that of female but with rounded process on proximal inner margin of second segment.

Legs 1–4 similar to those of female except for sexual dimorphism in third segment of endopod of leg 1 (Fig. 46d) where there is a terminal mammillate process in addition to the spiniform process.

Leg 5 (Fig. 46e) with free segment 20 × 13.5 μm.

Asteropontius parvipalpus Stock, 1975a

Host: *Condylactis gigantea* (Weinberg).
Site: On tentacles.
Localities: Piscadera Bay, Curaçao (Stock, 1975a); Bahamas, Puerto Rico, Jamaica (present paper).
Notes: According to Stock (1975a) this copepod is a constant associate of *Condylactis gigantea* in Piscadera Bay, with infestation about 15 copepods per host. Length of ♀ 1.45 mm, ♂ 0.96 mm.

New records (all from *Condylactis gigantea*): Jamaica: 17 ♀♀, 12 ♂♂, 3 copepodids from 6 hosts, in 1 m, Drunken Man's Cay, 28 August 1959. Bahamas: 7 ♀♀, 3 ♂♂ from 1 host, in 1 m, off Lerner Marine Laboratory, North Bimini, 1 June 1959; 10 ♀♀, 3 ♂♂ from 1 host, in 1 m, off Lerner Marine Laboratory, North Bimini, 3 June 1959; 125 ♀♀, 60 ♂♂, 19 copepodids from 30 hosts, northern end of Pigeon Cay, Bimini, 4 June 1959. Puerto Rico: 1 ♀, 2 ♂♂ from 3 hosts, in 0.5 m, near Magüeyes Island, southwestern Puerto Rico, 31 July 1959; 1 ♂, 2 copepodids from 1 host, on mangrove, southwestern corner of Mata Flores Island, southwestern Puerto Rico, 4 August 1959; 3 ♀♀, 4 ♂♂, 1 copepodid from 1 host, in 1 m, Laurel Reef, southwestern Puerto Rico, 13 August 1959; 3 ♀♀ from 1 host, on mangrove, reef called Mata Cagada, near La Parguera, southwestern Puerto Rico, 20 August 1959; 1 ♀, 1 copepodid from 1 host, in 0.5 m, Enrique Reef, near La Parguera, southwestern Puerto Rico, 17 August 1959; 1 ♀, 1 ♂ from 1 host, in 1 m, La Pelota, near La Cueva, between La Parguera and Cabo Rojo, southwestern Puerto Rico, 14 August 1959; 8 ♀♀, 7 ♂♂, 20 copepodids from 2 hosts, in 1 m, western shore of Cabo Rojo, southwestern Puerto Rico, 24 August 1959.

Asteropontius ungellatus Stock, 1975a

Host: *Stichodactyla* (= *Stoichactis*) *helianthus* (Ellis) [= *S. anemone* (Ellis)].
Site: Not given.
Locality: Near La Parguera, southwestern Puerto Rico (Stock, 1975a).
Notes: Length of ♀ 1.35 mm, ♂ 0.97 mm.

Host: *Phymanthus crucifer* (Lesueur).
Site: Not given.
Locality: Near La Parguera, southwestern Puerto Rico (Stock, 1975a).

Family Dinopontiidae Murnane, 1967
Genus *Dinopontius* Stock, 1960

Dinopontius acuticauda Stock, 1960

Host: *Anemonia sulcata* (Pennant).
Site: In washings.
Locality: Banyuls, France.
Notes: Length of ♀ 1.30–1.35 mm, ♂ 1.15 mm.

COPEPODS OF UNCERTAIN POSITION

Family Antheacheridae M. Sars, 1870
= Staurosomidae Zulueta, 1911

Okada (1927) regarded *Staurosoma* as extremely close to the Chondracanthidae.

For a discussion of possible affinities between the Antheacheridae and the Corallovexiidae Stock, 1975 (parasites of West Indian corals) see Stock (1975b).

Ho (1981) suggested that the Antheacheridae should probably include not only *Antheacheres*, *Gastroecus*, and *Staurosoma*, genera living in actiniarians, but also *Akessonia*, *Coelotrophus*, and *Siphonobius*, genera parasitizing sipunculans.

Genus *Antheacheres* M. Sars, 1857 (also 1870)

Antheacheres duebeni M. Sars, 1857 (also 1870)

Host: *Bolocera tuediae* (Johnston) [= *Anthea tuediae* (Johnston)]. [For synonymy see Andres (1883, p. 213) and Carlgren (1949, p. 54)].
Site: In galls formed from mesentery walls (Vader, 1970a).
Localities: Near Bergen and Dröbak, Norway (M. Sars, 1857, 1870); Gullmarfjorden, western Sweden (Bresciani and Lützen, 1962; Theel, 1907); Korsfjorden, western Norway (Vader, 1970a); northern Norway (Vader, 1975).
Notes: Length of ♀ up to 25 mm, ♂ up to 10 mm (Vader, 1970a). Electron micrograph of cuticle (Bresciani and Lützen, 1972).
Citation: Bouligand (1966). See also Dueben (1844, 1847) who first noticed a parasitic copepod in *Bolocera* in Norway (Vader, 1970a).

Genus *Gastroecus* Hansen, 1887
= *Parastaurosoma* Avdeev and Avdeev, 1975
Gastroecus arcticus Hansen, 1887
= *Parastaurosoma kamchaticum* Avdeev and Avdeev, 1975

Host: *Anthea* sp. [Vader (1970b) suggested that the host may actually have been *Actinostola spetsbergensis* Carlgren].
Site: Internal, exact location not given.
Locality: Kara Sea (Hansen, 1887).
Notes: Length of ♀ 17.5 mm, ♂ 2.6 mm. According to Vader (1970b) this copepod may belong to the genus *Antheacheres*, or more likely to *Staurosoma*, and is clearly a valid species.

Host: *Actinia equina* (L.), Actiniaria sp. I, Actiniaria sp. II.
Site: Gastral cavity.
Locality: Northwestern Pacific Ocean (Avdeev and Avdeev, 1975).
Notes: Length of ♀ 2.3–3.0 cm, ♂ 0.37 mm.

Gastroecus chukotensis Avdeev and Avdeev, 1978
= *Parastaurosoma arcticum* Avdeev and Avdeev, 1975

Host: *Bunodactis stella* Verrill.
Site: Gastral cavity.
Locality: Sea of Okhotsk (Avdeev and Avdeev, 1975).
Notes: Length of ♀ 1.2–2.2 cm.

Gastroecus okadai (Avdeev and Avdeev, 1975)
= *Parastaurosoma okadai* Avdeev and Avdeev, 1975

Host: *Tealia felina* (Cuvier).
Site: Gastral cavity.
Locality: Pacific coast of Japan (Avdeev and Avdeev, 1975).
Notes: Length of ♀ 1.1–2.0 cm, ♂ up to 5 mm.

Gastroecus caulleryi (Okada, 1927)
= *Staurosoma caulleryi* Okada, 1927
= *Parastaurosoma caulleryi* (Avdeev and Avdeev, 1975)

Host: *Nemanthus nitidus* (Wassilieff) (= *Sagartia nitida* Wassilieff).
Site: Internal, exact site not given.
Locality: Misaki, Japan (Okada, 1927).
Notes: Length of ♀ 13 mm, ♂ about 1 mm.
Citations: Bouligand (1966), Avdeev and Avdeev (1975, 1978).

Host: *Nemanthus annamensis* Carlgren.
Site: Coelenteric cavity.
Locality: Bay of Nhatrang, S. Annam; Réam, Cambodia.
Notes: Vader (1970b) has suggested that the copepods seen by Carlgren (1943) in *Nemanthus* may have been *Staurosoma caulleryi* (Okada), a copepod placed in this review in the genus *Gastroecus*, following Avdeev and Avdeev (1978). Vader pointed out that *Sagartia nitida* Wassilieff, the host of Okada's species, is the type-species of the genus *Nemanthus* Carlgren (cf. Carlgren, 1949, p. 110).

Genus *Staurosoma* Will, 1844
Staurosoma parasiticum Will, 1844

Host: *Anemonia sulcata* (Pennant) [= *Actinia viridis* (Forskål)]. For synonymy see Andres (1883, p. 197).
Site: In galls.
Localities: Mediterranean coast (Will, 1844); Channel coast of France and Gulf of Marseille (Caullery and Mesnil, 1902); Trieste, Italy (Heller, 1866; Stossich, 1880). (Host not given by Stossich but presumably *Anemonia sulcata*).
Notes: Length of ♀ up to 25 mm, ♂ 2 mm.
Citations: Zulueta (1911), Bouligand (1966).

Host: *Anthopleura stellula* Ehrenberg.
Site: In capsules formed by a mesenteric partition.
Locality: Gulf of Aqaba, Red Sea (Schmidt, 1970; Laubier and Schmidt, 1971).
Notes: Length of ♀ 5 mm (Laubier and Schmidt, 1971).

Undetermined species, probably Antheacheridae

Host: *Actinostola intermedia* Carlgren.
Site: In mesenteries.
Locality: Cape St. Vincent, Straits of Magellan (Carlgren, 1899).
Notes: "In den Mesenterien kommt eine parasitische Crustacee in verschiedenen Stadien vor. Die Grösseren waren etwa 1,5 cm lange Weibchen mit langen Eierschnüren. Die Parasiten wandern als junge Individuen in den coelenterischen Raum ein und setzen sich dort an den Mesenterien fest. Sobald die jungen Parasiten sich durch die Mesenterien

durchzubrechen versuchen, bildet das Mesenterium rings um die Parasiten eine Blindtasche, die auf der der Parasiten-Anheftung entgegengesetzten Seite zu liegen kommt. Mit dem Zuwachs des Parasiten wird die Blindtasche grösser und grösser. Die Mesenterien, welche ältere Parasiten enthalten, tragen also auf der einen Seite einen Blindsack, in dem der Parasit liegt, während der schmale, von dem jungen Parasiten gebildete Eingang dieses Sackes auf der anderen Seite sich befindet. Die älteren Parasiten können durch den schmalen Eingang des Sackes nicht mehr in den Gastrovaculärraum hineinkommen, sondern sind ganz und gar an die Blindtasche gefesselt." (Carlgren, 1899, pp. 32 and 33). In later papers Carlgren (1902, 1913, 1921, 1927) referred to copepods found in *Actinostola* as probably *Antheacheres* or a related genus.

Host: *Actinostola spetsbergensis* Carlgren.
Site: Internal.
Localities: 74°48'N, 20°54'E, off northern Norway (Carlgren, 1902); ca. 76°40'N, 84–88°W, north of Devon Island, northern Canada (Carlgren, 1913); Arctic, but no detailed localities given (Carlgren, 1921).
Notes: "In dem Innern fand ich eine parasitische Crustacee von ähnlichen Aussehen wie die in *A. intermedia* aus dem Antarctis erwähnten" (Carlgren, 1902, p. 47); "As I have mentioned (1902, p. 47) a parasitic Crustacean, probably *Antheachares dübenii* [sic], sometimes appears in the mesenteries" (Carlgren, 1921, p. 226).

Host: *Actinostola intermedia* Carlgren (= *Actinostola chilensis* McMurrich). See Carlgren (1927).
Site: In mesenteries.
Locality: Calbuco, Chile (Carlgren, 1927).
Notes: "In the mesenteries of the specimen I have observed the same parasitic Crustacean (probably in *Antheacheres* or related to this genus) which I have found in *intermedia* (Carlgren, 1899, p. 32)." [Carlgren (1927, pp. 60 and 61)].

Host: *Nemanthus annamensis* Carlgren.
Site: Coelenteron.
Localities: Anemones from Nhatrang, Viet Nam, and Réam, Cambodia, but exact localities of parasitized specimens not given (Carlgren, 1943).
Notes: "The coelenteron is often inhabited by a parasitic Copepod, probably belonging to *Antheacheres* or a nearly allied genus. There was usually one, sometimes 2, in each individual." (Carlgren, 1943, p. 38). The parasite from *Nemanthus annamensis* may have been *Gastroecus caulleryi* (Okada), as noted by Vader (1970b).

Host: *Nemanthus nitidus* (Wassilieff) (= *Sagartia nitida* Wassilieff).
Site: In the mesoglea of the septa.
Locality: Sagami Bay, Japan.
Notes: "Bei vielen Exemplaren dieser Actinie habe ich parasitische Copepoden beobachtet, die im Mesoderm der Septen als eine Anschwellung

zu sehen sind. Es treten dabei zwei Arten von parasitischen Copepoden auf. Die eine Art ist klein (3 mm), hat ovalen Körper, die Eiersäckchen liegen diesen dicht an; die andere ist grösser (ca. 1 cm), hat eckigen Körper und die Eiersäcke umwickeln in Gestalt von langen Schläuchen den ganzen Körper." (Wassilieff, 1908, p. 34). These copepods are from the same host and geographical area as the type material of *Gastroecus caulleryi* (Okada, 1927) and may well have been that species.

Family Mesoglicolidae Zulueta, 1911

The position of this monogeneric family is uncertain. Quidor (1936) thought it to be perhaps related to the Ascidicolidae. Bouligand (1966) recognized a resemblance of this family to the Lamippidae, copepods endoparasitic in octocorals.

Genus *Mesoglicola* Quidor, 1906

Mesoglicola delagei Quidor, 1906

Host: *Corynactis viridis* Allman.
Site: In the mesoglea (Quidor, 1906); in galls (Haefelfinger and Laubier, 1965).
Localities: Roscoff, northern France (Quidor, 1906, 1922, 1936; Taton, 1934); near Banyuls-sur-Mer, southern France (Haefelfinger and Laubier, 1965; Haefelfinger, reported in Laubier, 1966).
Notes: Length of ♀ about 7 mm, ♂ nearly one-half as long (Taton, 1934).

DISCUSSION

Host specificity

On the basis of available information, copepods living in association with sea anemones are generally specific in their choice of hosts. Among the copepods reported from sea anemones, three-fourths of the species (31) are restricted to one host species. Ten species of copepods occur on two species of sea anemones. One copepod, *Doridicola actiniae*, occurs on three hosts (*Actinia cari*, *Actinia equina*, and *Anemonia sulcata*). Certain records, however, involve only one or two specimens of a copepod on a host species. Such small numbers suggest that their presence may be accidental. This interpretation may be made in the case of only 1 ♀ and 1 ♂ *Doridicola magnificus* from one *Cryptodendrum adhaesivum*. The normal host of this copepod seems to be *Stichodactyla gigantea*, from which many specimens have been recovered (Humes, 1964).

In *Paranthessius anemoniae* in Northern Ireland, Briggs (1976) demonstrated a distinct preference for *Anemonia sulcata*, its natural host, over *Actinia equina* and *Metridium senile*. In artificial infection experiments, however, the copepod developed a degree of acclimation to the alien *Actinia equina*. The experimental tolerance of *Actinia* for *Paranthessius* raises the question as to why the copepod is not found in nature on *Actinia*. Briggs points out possible reasons: 1) *Actinia* usually lives on more exposed shores in turbulent waters less favorable for larval settlement and at higher levels on the shoreline subject to desiccation, salinity extremes, and wide temperature fluctuations, while *Anemonia* prefers calm sheltered waters; 2) *Anemonia* has long nonretractable tentacles hanging down over the column where the copepod normally lives and offering protection for the copepod, while *Actinia* has shorter tentacles which are often completely withdrawn.

Host specificity may extend to varieties within a species. The corallimorphian *Corynactis viridis* at Roscoff has several color forms whose populations in general do not mix. *Mesoglicola delagei* occurs in 50 per cent of the red varieties, but much less often in the green varieties (Taton, 1934). On the other hand, Lönning and Vader (personal communication) found that species of *Metaxymolgus* on Californian sea anemones showed no necessity for acclimation to a particular host individual or colony, and that intraspecific transfers always succeeded without problems.

Multiple associations with one host

Most collections of copepods from sea anemones are from an unspecified number of hosts or a group of hosts examined together. In several instances where hosts have been examined individually more than one species of copepod occurs on a single sea anemone. In the New Caledonian material there are seven examples of multiple associations with a single actiniarian. In *Heteractis crispa*, each of two hosts had three species of copepods (*Doridicola cylichnophorus, Doridicola paterellis*, and *Doridicola scyphulanus*), one host had four species of copepods (*D. cylichnophorus, D. paterellis, D. scyphulanus*, and *Notoxynus crinitus*), and one host had five species of copepods (*D. cylichnophorus, D. paterellis, D. scyphulanus, Doridicola dunnae*, and *N. crinitus*). One *Entacmaea quadricolor* had four species of copepods (*Doridicola caelatus, Doridicola hispidulus, Doridicola penicillatus*, and *Verutipes laticeps*). One *Cryptodendrum adhaesivum* had two species of copepods (*Lambanetes gemmulatus* and *Doridicola magnificus*). One *Stichodactyla gigantea* had two species of copepods (*Metaxymolgus cuspis* and *Lambanetes. stichodactylae*). Species occurring together on one host (as many as five) may have specific microhabitats on the host, may use different food material, or may have different behavior patterns or other barriers that prevent competition and allow their coexistence on the sea anemone. Little is known in this respect, although Lönning and Vader (personal communication) have recently made interesting observations of *Metaxymolgus* species on Californian sea anemones.

Food and feeding

The nature of the food of copepod associates is mostly unknown. In relatively unmodified external associates like *Paranthessius anemoniae* histochemical tests indicate that the gut contains material of a mucopolysaccharide nature as would be expected from a mucus feeder (Briggs, 1977a). Briggs thinks it likely that *Paranthessius* collects these substances as it "scrapes" the surface of the anemone and that such substances may render the particular host more attractive to the copepod.

Antheacheres duebeni has no digestive system and thus must derive its nourishment entirely through diffusion through the exceptionally thin body wall (Vader, 1970a; Bresciani, 1968). A mouth in *Staurosoma parasiticum* has not been identified (Laubier and Schmidt, 1971) and presumably this copepod also feeds by diffusion.

Color

While the color of many copepods associated with sea anemones is unknown or only superficially described, it appears that certain copepods take on the color of their host. During experimental infections of *Actinia equina* with *Paranthessius anemoniae*, after about three days on a red or green *Actinia*, the copepods appeared to resemble the color of this new host (Briggs, 1976). Copepods placed in colored dishes did not change color. The "red" copepods showed colored droplets under the cuticle, in the ovarian tubules, and in the gut. In addition, ovigerous "red" copepods laid red eggs which gave rise to red nauplii. Briggs thought it likely that xanthophyll esters responsible for coloration in *Actinia* are ingested by *Paranthessius*, absorbed, and transported to the eggs and subcuticular region without being chemically degraded.

Experiments by Lönning and Vader (personal communication) indicate that when starved individuals of *Metaxymolgus confinis* are placed on an *Anthopleura elegantissima* with tentacles of different color shades, the "digestive cross" of the copepods assumes the color of the host after 4–8 hours.

Effects on the host

Those copepods which move over the surface of the host seem not to produce marked responses on the part of the sea anemone. However, gallicolous copepods like *Staurosoma*, *Mesoglicola*, and *Antheacheres* live in galls or cysts which can be attributed to a specialized tissue response on the part of the anemone (Gotto, 1979). *Mesoglicola delagei* lives within large tumors or galls in *Corynactis viridis*. These galls are prominent externally as shown in photographs by Haefelfinger and Laubier (1965). *Staurosoma parasiticum* in *Anthopleura stellula* lies in mesenteric capsules with thick walls, filled with debris, and open on the equator (Laubier and Schmidt, 1971). In the region of the capsule, the mesentery loses its filaments and part of the retractor muscle. Infested *Anthopleura* cannot be distinguished externally from uninfested anemones. In *Antheacheres duebeni* the galls are formed from the mesentery walls of *Bolocera tuediae* and connected to the walls by a constricted base (Vader, 1970a). Up to twenty galls may occur in a clump like a bunch of grapes, although they apparently have no direct intercommunication. The galls, which may contain as many as twelve copepods, are closed and contain a clear fluid in which the copepods lie free.

Immunity to anemones

Copepod associates of actiniarians have an apparent immunity to the toxin-producing nematocysts usually found in the tentacles of these cnidarians. In experiments on copepod immunity to *Anemonia sulcata*, Briggs (1976) found that of several copepod species tested only *Paranthessius anemoniae* moved freely among the tentacles, while other copepods were either paralyzed and digested or showed avoidance reaction. Extract of tentacles of *Anemonia* had little effect on *Paranthessius* but the extract appeared to paralyze temporarily other species.

Development

In relatively unmodified copepods associated externally with sea anemones, the size of the more or less spherical egg varies with egg number. The smaller the number of eggs in an egg sac, the larger are the individual eggs (Gotto, 1979). In *Mesoglicola delagei*, two or three pairs of egg sacs may be found in the same gall in *Corynactis viridis*, corresponding to the same number of successive egg layings (Laubier, 1966).

The larval stages of *Paranthessius anemoniae*, an unmodified associate, consist of six nauplius stages and at least two copepodid stages (Briggs, 1977b). In *Mesoglicola delagei*, a modified gallicolous form, the larval stages are poorly known, but Taton (1934) described a first nauplius and two cyclopoid stages, A and B, before reaching the adult stage. Her observations indicated that the first nauplius penetrates the host and suggested that cyclopoid stage B may also be capable of this. Laubier (1966) was able to rear some metanauplius stages of *Mesoglicola*. According to this author the question of how the nauplii of *Mesoglicola* escape from the galls is unresolved.

In *Aspidomolgus stoichactinus* copepodids II, IV, and V have been described (Humes, 1969).

COPEPODS AND HOSTS

LIST OF COPEPODS AND THEIR ACTINIARIAN AND CORALLIMORPHARIAN HOSTS

Antheacheres duebeni
  Bolocera tuediae
Aspidomolgus stoichactinus
  Homosticanthus duerdeni (= *H. denticulosus*)
  Stichodactyla helianthus
Asterocheres scutatus
  Rhodactis rhodostoma
Asteropontius longipalpus
  Ricordea florida
Asteropontius parvipalpus
  Condylactis gigantea
Asteropontius ungellatus
  Phymanthus crucifer
  Stichodactyla helianthus
Dinopontius acuticauda
  Anemonia sulcata
Doridicola actiniae (= *Lichomolgus actiniae*, = *Lichomolgus anemoniae*)
  Actinia cari
  Actinia equina
  Anemonia sulcata
Doridicola antheae (perhaps = *Paranthessius anemoniae*)
  Anemonia sulcata
Doridicola caelatus
  Entacmaea quadricolor
Doridicola cylichnophorus
  Heteractis crispa
Doridicola dunnae
  Heteractis crispa
Doridicola gemmatus (= *Lichomolgus gemmatus*)
  Stichodactyla gigantea
Doridicola hispidulus
  Entacmaea quadricolor
Doridicola magnificus (= *Lichomolgus magnificus*)
  Cryptodendrum adhaesivum
  Stichodactyla gigantea
Doridicola paterellis
  Heteractis crispa

Doridicola penicillatus
  Entacmaea quadricolor
Doridicola scyphulanus
  Heteractis crispa
Doridicola titillans
  Condylactis gigantea
Gastroecus arcticus (= *Parastaurosoma kamchaticum*)
  Actinia equina
  Anthea sp.
Gastroecus caulleryi (= *Staurosoma caulleryi*, = *Parastaurosoma caulleryi*)
  Nemanthus annamensis
  Nemanthus nitidus
Gastroecus chukotensis (= *Parastaurosoma arcticum*)
  Bunodactis stella
Gastroecus okadai (= *Parastaurosoma okadai*)
  Tealia felina
Indomolgus panikkari (= *Lichomolgus panikkari*)
  Phytocoeteopsis ramunni
Lambanetes gemmulatus
  Cryptodendrum adhaesivum
Lambanetes stichodactylae
  Stichodactyla gigantea
  Stichodactyla haddoni
Laophonte adamsiae
  Adamsia palliata
Lichomolgus actiniae (= *Doridicola actiniae*)
Lichomolgus cuspis (= *Metaxymolgus cuspis*)
Lichomolgus gemmatus (= *Doridicola gemmatus*)
Lichomolgus magnificus (= *Doridicola magnificus*)
Lichomolgus myorae (= *Metaxymolgus myorae*)
Lichomolgus panikkari (= *Indomolgus pannikkari*)
Lichomolgus politus (= *Paramolgus politus*)
Lichomolgus simulans (= *Paramolgus simulans*)
Mesoglicola delagei
  Corynactis viridis
Metaxymolgus confinis
  Anthopleura elegantissima
  Anthopleura xanthogrammica
  Tealia piscivora
Metaxymolgus cuspis (= *Lichomolgus cuspis*)
  Heteractis magnifica (= *Radianthus ritteri*)
  Stichodactyla gigantea
Metaxymolgus myorae (= *Lichomolgus myorae*)
  Stichodactyla haddoni
Metaxymolgus pertinax
  Tealia coriacea

Tealia crassicornis
Metaxymolgus sunnivae
  Epiactis prolifera
  Tealia crassicornis
  Tealis lofotensis
Metaxymolgus turmalis
  Anthopleura artemisia
Notoxynus crinitus
  Heteractis crispa
Paramolgus antillianus
  Ricordea florida
Paramolgus politus (= *Lichomolgus politus*)
  Rhodactis rhodostoma
Paramolgus simulans (= *Lichomolgus simulans*)
  Rhodactis rhodostoma
Paranthessius anemoniae
  Anemonia sp.
  Anemonia sulcata
Parastaurosoma arcticum (= *Gastroecus chokotensis*)
Parastaurosoma caulleryi (= *Gastroecus caulleryi*)
Parastaurosoma kamchaticum (= *Gastroecus arcticus*)
Parastaurosoma okadai (= *Gastroecus okadai*)
Ridgewayia fosshageni
  Bartholomea annulata
Staurosoma caulleryi (= *Gastroecus caulleryi*)
Staurosoma parasiticum
  Anemonia sulcata
  Anthopleura stellula
Undetermined species
  Actinostola intermedia
  Actinostola spetsbergensis
  Nemanthus annamensis
  Nemanthus nitidus
Verutipes laticeps
  Entacmaea quadricolor

LIST OF ACTINIARIANS AND CORALLIMORPHARIANS AND THEIR ASSOCIATED COPEPODS

Actinia cari (= *Actinia concentrica*, **var.** *viridis*)
  Doridicola actiniae
Actinia concentrica, **var.** *viridis* (= *Actinia cari*)
Actinia equina
  Doridicola actiniae
  Gastroecus arcticus
Actinia viridis (= *Anemonia sulcata*)
Actinostola chilensis (= *Actinostola intermedia*)

Actinostola intermedia (= *Actinostola chilensis*)
  Undetermined species
Actinostola spetsbergensis
  Undetermined species
Actinostola spetsbergensis (= *Anthea* sp.)
Adamsia palliata
  Laophonte adamsiae
Anemonia sp.
  Paranthessius anemoniae
Anemonia sulcata (= *Anthea cereus,* = *Actinia viridis*)
  Dinopontius acuticauda
  Doridicola actiniae
  Doridicola antheae
  Paranthessius anemoniae
  Staurosoma parasiticum
Anthea cereus (= *Anemonia sulcata*)
Anthea sp. (perhaps = *Actinostola spetsbergensis*)
  Gastroecus arcticus
Anthea tuediae (= *Bolocera tuediae*)
Anthopleura artemisia
  Metaxymolgus turmalis
Anthopleura elegantissima
  Metaxymolgus confinis
Anthopleura stellula
  Staurosoma parasiticum
Anthopleura xanthogrammica
  Metaxymolgus confinis
Bartholomea annulata
  Ridgewayia fosshageni
Bolocera tuediae (= *Anthea tuediae*)
  Antheacheres duebeni
Bunodactis stella
  Gastroecus chukotensis
Condylactis gigantea
  Asteropontius parvipalpus
  Doridicola titillans
Corynactis viridis
  Mesoglicola delagei
Cryptodendrum adhaesivum
  Doridicola magnificus
  Lambanetes gemmulatus
Entacmaea quadricolor
  Doridicola caelatus
  Doridicola hispidulus
  Doridicola penicillatus
  Verutipes laticeps

Epiactis prolifera
  Metaxymolgus sunnivae
Heteractis crispa
  Doridicola cylichnophorus
  Doridicola dunnae
  Doridicola paterellis
  Doridicola scyphulanus
  Notoxynus crinitus
Heteractis magnifica
  Metaxymolgus cuspis
Homosticanthus duerdeni (= *H. denticulosus*)
  Aspidomolgus stoichactinus
Nemanthus annamensis
  Gastroecus caulleryi
  Undetermined species
Nemanthus nitidus (= *Sagartia nitida*)
  Gastroecus caulleryi
  Undetermined species
Phymanthus crucifer
  Asteropontius ungellatus
Phytocoeteopsis ramunni
  Indomolgus panikkari
Radianthus ritteri (= *Heteractis magnifica*)
Rhodactis rhodostoma
  Asterocheres scutatus
  Paramolgus politus
  Paramolgus simulans
Ricordea florida
  Asteropontius longipalpus
  Paramolgus antillianus
Sagartia nitida (= *Nemanthus nitidus*)
Stichodactyla helianthus (= *Stoichactis anemone*)
  Aspidomolgus stoichactinus
  Asteropontius ungellatus
Stichodactyla gigantea (= *Stoichactis giganteum*)
  Doridicola gemmatus
  Doridicola magnificus
  Lambanetes stichodactylae
  Metaxymolgus cuspis
Stichodactyla haddoni (= *Stoichactis haddoni*)
  Lambanetes stichodactylae
  Metaxymolgus myorae
Stoichactis giganteum (= *Stichodactyla gigantea*)
Stoichactis haddoni (= *Stichodactyla haddoni*)
Stoichactis helianthus (= *Stichodactyla anemone*)
Tealia coriacea

Metaxymolgus pertinax
Tealia crassicornis
  Metaxymolgus pertinax
  Metaxymolgus sunnivae
Tealia felina
  Gastroecus okadai
Tealia lofotensis
  Metaxymolgus sunnivae
Tealia piscivora
  Metaxymolgus confinis

COPEPODS WITHIN HOST FAMILIES

DISTRIBUTION OF COPEPODS WITHIN 11 FAMILIES OF ACTINIARIA AND CORALLIMORPHARIA

	No. of copepod species
Order Actiniaria	
Haliactiidae	
Phytocoeteopsis	1
Actiniidae	
Actinia	3
Anemonia	5
Anthopleura	4
Bolocera	1
Bunodactis	1
Entacmaea	4
Epiactis	1
Condylactis	2
Tealia	3
Thalassianthidae	
Cryptodendrum	1
Homostichanthidae	
Homostichanthus	1
Stichodactylidae	
Stichodactyla	7
Heteractis	6
Phymanthidae	
Phymanthus	1
Hormathiidae	
Adamsia	1
Aiptasiidae	
Bartholomea	1
Nemanthidae	
Nemanthus	1
Order Corallimorpharia	
Actinodiscidae	
Rhodactis	3
Corallimorphidae	
Corynactis	1
Ricordea	2

REFERENCES

Andres, A. 1883. Le attinie. *R. Accad. Lincei, ser. 3, Mem. Cl. Sci. Fis., Mat. e Nat.* 14: 1–460.

Avdeev, G. V. and V. V. Avdeev 1975. Copepods of the family Staurosomatidae—parasites of Actiniaria. *Biologiia Moria* 4: 3–12. (In Russian).

—— 1978. *Parastaurosoma* G. Avdeev et V. Avdeev, 1975—synonym of the genus *Gastroecus* Hansen, 1887 (Copepoda, Staurosomatidae). *Parasitologia* 12: 448–449. (In Russian).

Bocquet, C. and J. H. Stock 1959. Copépodes parasites d'invertébrés des côtes de la Manche. VI. Redescription de *Paranthessius anemoniae* Claus (Copepoda Cyclopoida), parasite d'*Anemonia sulcata* (Pennant). *Arch. Zool. Exp. Gen.* 98 (Notes et Revue no. 1): 43–53.

Boeck, A. 1859. Beskrivelse over tvende nye parasitiske Krebsdyr. *Forh. Vidensk.-Selsk. Christiania* 1859: 171–182.

Bouligand, Y. 1966. Recherches récentes sur les copépodes associés aux anthozoaires. In: W. J. Rees, ed., *The Cnidaria and Their Evolution*. Symposia Zool. Soc. London (Academic Press, London and New York) 16: 267–306.

Bresciani, J. 1968. Den cuticulare struktur hos parasitiske Crustaceer. *Tiedoksianto* 9: 32.

—— and J. Lützen 1962. Parasitic copepods from the west coast of Sweden including some new or little known species. *Vidensk. Medd. fra Dansk Naturh. Foren.* 124: 367–408.

—— 1972. The sexuality of *Aphanodomus* (parasitic copepod) and the phenomenon of cryptogonochorism. *Vidensk. Medd. fra Dansk Naturh. Foren.* 135: 7–20.

Briggs, R. P. 1973. *Lichomolgus actiniae* D. V.: an associated copepod new to Irish waters. *Irish Nat. Jour.* 17(12). 423.

—— 1976. Biology of *Paranthessius anemoniae* in association with anemone hosts. *J. Mar. Biol. Ass. U.K.* 56: 917–924.

—— 1977a. Structural observations on the alimentary canal of *Paranthessius anemoniae*, a copepod associate of the snakelocks anemone *Anemonia sulcata*. *J. Zool., London*, 182: 353–368.

—— 1977b. Larval stages of *Paranthessius anemoniae* Claus (Copepoda, Cyclopoida), an associate of the snakelocks anemone *Anemonia sulcata* (Pennant). *Crustaceana* 33: 249–358.

—— and R. V. Gotto 1973. A first record of *Lichomolgus actiniae* Della Valle, 1880 (Copepoda, Cyclopoida) in British Waters. *Crustaceana* 24(3): 336–337.

Carlgren, O. 1899. Zoantharien. *Hamburger Magalhaensische Sammelreise* 4: 1–48.

—— 1902. Die Actiniarien der Olga-Expedition. *Wissensch. Meeresunters., Abt. Helgoland*, neue Folge 5: 33–56.

—— 1913. Actiniaria. Rept. Second Norwegian Arctic Exped. "Fram": 3–6.

—— 1921. Actiniaria part I. *Danish Ingolf-Exped.* 5(9): 1–241.

—— 1927. Actiniaria and Zoantharia. *Further Zool. Res. Swedish Antarctic Exped.* 2(3): 1–102.

—— 1943. East-Asiatic Corallimorpharia and Actiniaria. *Kungl. Svenska Vetensk. Akad. Handl.* ser. 3, 20(6): 1–43.

—— 1949. A survey of the Ptychodactiaria, Corallimorpharia and Actiniaria. *Kungl. Svenska Vetensk. Akad. Handl.*, ser. 4, 1(1): 1–121.

Carton, Y. 1963. Etude de la spécificité parasitaire chez *Lichomolgus actiniae* D. V. (Copépode Cyclopoïde). *C. R. Acad. Sci., Paris* 256: 1148–1150.

Caullery, M. and F. Mesnil 1902. Sur *Staurosoma parasiticum* Will, copépode gallicole, parasite d'une actinie. *C. R. Acad. Sci., Paris* 134: 1314–1317.

Claus, C. 1889. Uber neue oder wenig bekannte halbparasitische Copepoden, insbesondere der Lichomolgiden- und Ascomyzontiden-Gruppe. *Arb. Zool. Inst. Univ. Wien und Zool. Stat. Triest* 8(3): 1–44.

Della Valle, A. 1880a. Sui coriceidi parassiti, e sull'anatomia del gen. *Lichomolgus*. *Mitt. Zool. Station Neapel* 2: 83–106.

—— 1880b. Sui coriceidi parassiti, e sull' anatomia del gen. *Lichomolgus*. *Atti. R. Accad. Lincei*, ser. 3, Mem. Cl. Sci. Fis., Mat. e Nat. 5: 107–124.

Dueben, M. W. von 1844. Om Norriges Hafsfauna. *Öfvers. K. VetenskAkad. Förh.* 1: 13–15.

—— 1847. (Demonstration) *Forh. Skand. Naturf. Möte* 4(1844): 280.

Dunn, D. F. 1981. The clownfish sea anemones: Stichodactylidae (Coelenterata:Actiniaria) and other sea anemones symbiotic with pomacentrid fishes. *Trans. Amer. Philos. Soc.* 71(2): 1–115.

Giesbrecht, W. 1899. Die Asterocheriden des Golfes von Neapel und der angrezenden Meeres-Abschnitte. *Fauna Flora Golfes Neapel* 25: 1–217.

Gnanamuthu, C. P. 1955. A new semi-parasitic copepod from an estuarine actiniarian of Madras. *Rec. Ind. Mus.* 52(1): 151–156.

Gooding, R. U. 1957. On some Copepoda from Plymouth, mainly associated with invertebrates, including three new species. *Jour. Mar. Biol. Assoc. U.K.* 36: 151–156.

Gotto, R. V. 1979. The association of copepods with marine invertebrates." *Adv. Mar. Biol.* 16: 1–109.

—— and R. P. Briggs 1972. *Paranthessius anemoniae* Claus: an associated copepod new to British and Irish waters. *Irish Nat. Jour.* 17(7): 243–244.

Graeffe, E. 1900. Uebersicht der Fauna des Golfes von Triest nebst Notizen über Vorkommen, Lebensweise, Erscheinungs- und Laichzeit der einzelnen Arten. *Arb. Zool Inst. Univ. Wien und Zool. Stat. Triest* 13(1): 33–80.

Greenwood, J. G. 1971 A new species of *Lichomolgus* (Copepoda, Cyclopoida) from an actiniarian host. *Crustaceana* 21(3): 298–306.

Gurney, R. 1927. Zoological results of the Cambridge Expedition to the Suez Canal, 1924. XXXIII. Report on the Crustacea: Copepoda (littoral and semiparasitic). *Trans. Zool. Soc. London*, 22(4): 451–577.

Haefelfinger, H. R. and L. Laubier 1965. Découverte en Méditerranée occidentale de *Mesoglicola delagei* Quidor, copépode parasite d'actinies. *Crustaceana* 9: 210–212.

Hansen, H. J. 1887. Oversigt over de paa Dijmphna-Togtet indsamlede Krebsdyr. In: C. F. Lütken, ed., *Dijmphna-Togtets zoologisk botaniske Udbytte*. Kjøbenhavn Universitets Zoologiske Museum, Kjøbenhavn, 183–286.

Heller, C. 1866. Carcinologische Beiträge zur Fauna des adriatischen Meeres. *Verhandl. kaiserl.-konigl. Zool.-bot. Gesellsch. Wien* 16: 723–760.

Ho, J.-S. 1981. Parasitic Copepoda of gastropods from the Sea of Japan. *Ann. Rep. Sado Mar. Biol. Stat., Niigata Univ.*, 11:23–41.

——, F. Katsumi, and Y. Honma 1981. *Coelotrophus nudus* gen. et sp. nov., an endoparasitic copepod causing sterility in a sipunculan *Phascolosoma scolops* (Selenka and De Man) from Sado Island, Japan. *Parasitology* 82:481–488.

Humes, A. G. 1959. Copépodes parasites de mollusques à Madagascar. *Mem. Inst. Sci. Madagascar*, 1958, ser. F, 2: 285–342.

—— 1964. New species of *Lichomolgus* (Copepoda, Cyclopoida) from sea anemones and nudibranchs in Madagascar. *Cahiers ORSTOM Océanogr.*, no. 6 (ser. Nosy Bé II), 59–130.

—— 1969. *Aspidomolgus stoichactinus* n. gen., n. sp. (Copepoda, Cyclopodia) associated with an actiniarian in the West Indies. *Crustaceana* 16(3): 225–242.

—— 1970. A census of copepods associated with marine invertebrates in a tropical locality. *Jour. Parasitol.* 56: 160–161.

—— 1975. Cyclopoid copepods (Lichomolgidae) associated with alcyonaceans in New Caledonia. *Smithsonian Contr. Zool.*, 191: 1–27.

—— 1978. Lichomolgid copepods (Cyclopoida), with two new species of *Doridicola*, from sea pens (Pennatulacea) in Madagascar. *Trans. Amer. Micros. Soc.* 97(4): 524–539.

—— and R. U. Gooding 1964. A method for studying the external anatomy of copepods. *Crustaceana* 6(3): 238–240.

—— and J.-S. Ho 1966. New lichomolgid copepods (Cyclopoida) from zoanthid coelenterates in Madagascar. *Cahiers ORSTOM Océanogr.* 4(2): 3–47.

—— 1967. Two new species of *Lichomolgus* (Copepoda, Cyclopoida) from an actiniarian in Madagascar. *Cahiers ORSTOM Océanogr.* 5(1): 3–21.

—— and W. L. Smith 1974. *Ridgewayia fosshageni* n. sp. (Copepoda, Calanoida) associated with an actiniarian in Panama, with observations on the nature of the association. *Carib. Jour. Sci.* 14(3–4): 125–139.

—— and J. H. Stock 1973. A revision of the family Lichomolgidae Kossmann, 1877, cyclopoid copepods mainly associated with marine invertebrates. *Smithsonian Contr. Zool.*, 127: 1–368.

Kabata, Z. 1979. Parasitic Copepoda of British fishes. *The Ray Soc., London*, no. 152, 1–468.

Kossmann, R. 1877 Entomostraca (1. Theil: Lichomolgidae). In: *Zoologische Ergebnisse einer im Auftrage der Königlichen Academie der Wissenschaften zu Berlin ausgeführten Reise in die Küstengebiete des Rothen Meeres*, erste Hälfte, IV, 1–24.

Laubier, L. 1966. Le coralligène des Albères. Monographie biocénotique. *Ann. Inst. Océanogr.*, new series, 43(2): 137–316.

—— and H. Schmidt 1971. Le genre *Staurosoma* Will (copépode parasite d'actinies) en Mer Rouge. *Crustaceana* 21(1): 52–56.

Leydig, F. 1853. Zoologische Notizen 1. Neuer Schmarotzerkrebse auf einem Weichthier. *Zeitschr. Wiss. Zool.* 4(3/4): 377–382.

Murnane, J. P. 1967. A new family and genus of cyclopoid copepods associated with the sponge *Cliona celata* (Grant) in Massachusetts. *Crustaceana* 12(3): 225–232.

Okada, Y. K. 1927. *Staurosoma*, copépode parasite d'une actinie: description de *Staurosoma caulleryi* sp. n. *Annot. Zool. Jap.* 11: 173–183.

Philippi, A. 1840. Zoologische Bemerkungen. *Arch. Naturg.* (6)1: 181–195.

Quidor, A. 1906. Sur *Mesoglicola delagei* (n. g., n. s.) parasite de *Corynactis viridis*. C. R. Acad. Sci., Paris, 143: 613–615.

—— 1922. Sur '*Mesoglicola delagei*' Quidor et son hôte. *Ann. Sci. Nat., Zool.*, ser. 10, 5: 77–81.

—— 1936. Sur *Leposphilus labrei* et *Mesoglicola delagei. Livre Jubil. Eugène Louis Bouvier*, 67–71. Firmin-Didot, Paris.

Raibaut, A. 1966. *Laophonte adamsiae* n. sp. (Copepoda, Harpacticoida) inquilin d'une actinie, *Adamsia palliata* (Bohadsch). *Crustaceana* 11(2): 123–128.

Ridley, H. N. 1879. On a new copepod of the genus *Doridicola. Ann. Mag. Nat. Hist.*, ser. 5, 4: 458.

Sars, G. O. 1903. An account of the Crustacea of Norway with short descriptions and figures of all the species. V. Copepoda Harpacticoida. Parts I and II Misophriidae, Longipediidae, Cerviniidae, Ectinosomidae (part), pp. 1–28. Bergen Museum, Norway.

Sars, M. 1857. *Antheacheres Duebenii* Sars, nov. gen. & spec. *Forh. Skand. Naturf. Möte* 7(1856): 175–181.

—— 1870. Bidrag til Kundskab om Christianiafjordens Fauna. *Nyt Magazin for Naturvidensk.* 17: 113–226.

Schmidt, H. 1970. *Anthopleura stellula* (Actiniaria, Actiniidae) and its reproduction by transverse fission. *Mar. Biol.* 5: 245–255.

—— 1972. Prodromus zu einer Monographie der mediterranen Akinien. *Zoologica* 42(2)(121): 1–146.

Scott, T. 1905. On some new and rare Crustacea from the Scottish seas. *23rd Ann. Rept. Fish. Bd. Scotland*, 1904, III: 141–153.

Shen, C.-J. and F.-S. Lee 1966. On the estuarine copepods of Chaikiang River, Kwantung Province. *Acta Zootaxonomica Sinica* 3(3): 212–233.

Stock, J. H. 1959. Copepoda associated with Neapolitan invertebrates. *Pubbl. Staz. Zool. Napoli* 31(1): 59–75.

—— 1960. Sur quelques copépodes associés aux invertébrés des côtes du Roussillon. *Crustaceana* 1(3): 218–257.

—— 1966. Copepoda associated with invertebrates from the Gulf of Aquaba. 1. The genus *Asterocheres* Boeck, 1859 (Cyclopoida, Asterocheridae). *Proc. Koninkl. Nederl. Akad. Wetensch., Amsterdam*, ser. C, 69(2): 204–210.

—— 1967. Copepoda associated with invertebrates from the Gulf of Aqaba. 4. Two new Lichomolgidae associated with Crinoida." *Proc. Koninkl. Nederl. Akad. Wetensch.—Amsterdam,* ser. C, 70(5): 569–578.

—— 1975a. Copepoda associated with West Indian Actiniaria and Corallimorpharia." *Stud. Fauna Curaçao and other Carib. Is.* 48(161): 88–118.

—— 1975b. Corallovexiidae, a new family of transformed copepods endoparasitic in reef corals. *Stud. Fauna Curaçao and other Carib. Is.* 47: 1–45.

—— and G. Kleeton 1963. Copépodes associés aux invertébrés des côtes du Rousillon. 2.— Lichomolgidae ecto-associés d'octocoralliaires. *Vie et Milieu* 14(2): 245–261.

Stossich, M. 1880. Prospetto della fauna del mare Adriatico. Parte III. *Boll. Soc. Adriatica Sci. Nat. Trieste* 6: 178–271.

Taton, H. 1934. Contribution à l'étude du copépode gallicole *Mesoglicola delagei* Quidor. *Trav. Stat. Biol. Roscoff* 12: 53–68.

Theel, H. 1907. Om utvecklingen af Sveriges zoologiska hafsstation Kristineberg och om djur-lifvet i angränsande haf och fjordar. *Ark. Zool.* 4(5): 1–136.

Thompson, I. C. and A. Scott 1903. Report of the Copepoda collected by Professor Herdman, at Ceylon, in 1902. *Rept. Govt. Ceylon Pearl Oyster Fish. Gulf of Manaar*, suppl. repts., 7: 227–307.

Ummerkutty, A. N. P. 1962. Studies on Indian copepods 5. On eleven new species of marine cyclopoid copepods from the south-east coast of India. *J. Mar. Biol. Ass. India*, 1961, 3(1 & 2): 19–69.

Vader, W. 1970a. *Antheacheres duebeni* M. Sars, a copepod parasitic in the sea anemone *Bolocera tuediae* (Johnston). *Sarsia* 43: 99–106.

—— 1970b. On the occurrence of a gall-forming copepod in *Actinostola* spp. (Anthozoa). *Sarsia* 43: 107–110.

—— 1975. The sea anemone *Bolocera tuediae* and its copepod parasite, *Antheacheres duebeni*, in northern Norway. *Astarte* 8: 37–39.

Wassilieff, A. 1908. Japanische Actinien. In: Beiträge zur Naturgeschichte Ostasiens, F. Doflein ed. *Abhandl. Mat.-Phys. Kl. K. Bayer, Akad. Wiss.*, suppl. 1, 2: 1–52.

Will, F. 1844. Uber *Staurosoma*, einen in den Aktinien lebenden Schmarotzer. *Arch. Naturg.*, Jahrg. 10, 1: 337–344.

Wilson, M. S. 1958. A review of the copepod genus *Ridgewayia* (Calanoida) with descriptions of new species from the Dry Tortugas, Florida. *Proc. U.S. Nat. Mus.* 108: 137–179.

Zulueta, A. de 1911. Los copépodos parásitos de los celentéreos. *Mem. Real Soc. Esp. Hist. Nat.* 7: 5–58.

INDEX

www.ingramcontent.com/pod-product-compliance
Lightning Source LLC
Chambersburg PA
CBHW061756260326
41914CB00006B/1123